Principles of Biochemistry

Principles of Biochemistry

Edited by
Tate Campbell

Larsen & Keller
www.larsen-keller.com

Principles of Biochemistry
Edited by Tate Campbell
ISBN: 978-1-63549-007-7 (Hardback)

🞃 Larsen & Keller

Published by Larsen and Keller Education,
5 Penn Plaza,
19th Floor,
New York, NY 10001, USA

Cataloging-in-Publication Data

Principles of Biochemistry / edited by Tate Campbell.
      p. cm.
Includes bibliographical references and index.
ISBN 978-1-63549-007-7
1.Biochemistry--Textbooks. I. Campbell, Tate.
QH345 .E87 2017
572.3--dc23

The publisher's policy is to use permanent paper from mills that operate a sustainable forestry policy. Furthermore, the publisher ensures that the text paper and cover boards used have met acceptable environmental accreditation standards.

Printed and bound in the United States of America.

For more information regarding Larsen and Keller Education and its products, please visit the publisher's website www.larsen-keller.com

Table of Contents

Preface

This book attempts to understand the multiple branches that fall under the discipline of biochemistry and how such concepts have practical applications. It provides detailed information about the basic principles of this field. Biochemistry as a field of science deals with the study of all the chemical processes happening inside living organisms. It is often referred to as biological chemistry. Some of the diverse topics covered in this text address the varied methodologies that comprise this subject. This textbook is an essential guide for both graduates and post-graduates who wish to pursue biochemistry further.

Given below is the chapter wise description of the book:

Chapter 1- Biochemistry is a rapidly growing field. The progress of this field has resulted in a better understanding of living processes across all forms of life. This chapter is written to provide students a general introduction about the subject. It aims to emphasize on the basics of the subject along with understanding its importance in the modern times.

Chapter 2- This chapter will deal in the major and minor categories of biological macromolecules. The major macromolecules discussed here are carbohydrates, nucleic acids, proteins and lipids. Many forms and functions of lipids are also discussed in the chapter like lipid microdomain, lipidomics, phenolic lipids and protein-lipid interactions. The chapter is written in a comprehensive manner to provide easy and simplistic explanations of concepts.

Chapter 3- This chapter discusses enzyme secretion and formation in living beings and also sheds light on the enzyme related activity. It provides detailed information about proteolysis and enzymes to the students. It discusses their structure, mechanism, etymology, inhibition and history. Students will get in-depth knowledge about enzymes from this chapter.

Chapter 4- Metabolism plays an important role in biochemistry. Many experiments and theories in biochemistry are conducted by studying metabolism. This chapter will help students understand the simple as well as complex concepts of metabolism in biochemistry. It will also provide information about the genetic code and what it is made of.

Chapter 5- This chapter will discuss all the allied fields of biochemistry. The major topics discussed here are molecular biology, cell biology, biotechnology and bioluminescence. All the topics mentioned in the chapter are rapidly growing branches of biochemistry and are, therefore, an important part of this field.

Chapter 6- This chapter provides thorough knowledge about the history, evolution and current importance of biochemistry. It provides information about enzymes and metabolism. The chapter aims to elaborate the evolution of biochemistry since its early stages to becoming a specialized field in the modern times.

Indeed, my job was extremely crucial and challenging as I had to ensure that every chapter is informative and structured in a student-friendly manner. I am thankful for the support provided by my family and colleagues during the completion of this book.

Editor

Introduction to Biochemistry

Biochemistry is a rapidly growing field. The progress of this field has resulted in a better understanding of living processes across all forms of life. This chapter is written to provide students a general introduction about the subject. It aims to emphasize on the basics of the subject along with understanding its importance in the modern times.

Biochemistry, sometimes called biological chemistry, is the study of chemical processes within and relating to living organisms. By controlling information flow through biochemical signaling and the flow of chemical energy through metabolism, biochemical processes give rise to the complexity of life. Over the last decades of the 20th century, biochemistry has become so successful at explaining living processes that now almost all areas of the life sciences from botany to medicine to genetics are engaged in biochemical research. Today, the main focus of pure biochemistry is on understanding how biological molecules give rise to the processes that occur within living cells, which in turn relates greatly to the study and understanding of tissues, organs, and whole organisms—that is, all of biology.

Biochemistry is closely related to molecular biology, the study of the molecular mechanisms by which genetic information encoded in DNA is able to result in the processes of life. Depending on the exact definition of the terms used, molecular biology can be thought of as a branch of biochemistry, or biochemistry as a tool with which to investigate and study molecular biology.

Much of biochemistry deals with the structures, functions and interactions of biological macromolecules, such as proteins, nucleic acids, carbohydrates and lipids, which provide the structure of cells and perform many of the functions associated with life. The chemistry of the cell also depends on the reactions of smaller molecules and ions. These can be inorganic, for example water and metal ions, or organic, for example the amino acids, which are used to synthesize proteins. The mechanisms by which cells harness energy from their environment via chemical reactions are known as metabolism. The findings of biochemistry are applied primarily in medicine, nutrition, and agriculture. In medicine, biochemists investigate the causes and cures of diseases. In nutrition, they study how to maintain health and study the effects of nutritional deficiencies. In agriculture, biochemists investigate soil and fertilizers, and try to discover ways to improve crop cultivation, crop storage and pest control.

History

At its broadest definition, biochemistry can be seen as a study of the components and composition of living things and how they come together to become life, and the history of biochemistry may therefore go back as far as the ancient Greeks. However, biochemistry as a specific scientific discipline has its beginning some time in the 19th century, or a little earlier, depending on which aspect of biochemistry one is being focused on. Some argued that the beginning of biochemistry may have been the discovery of the first enzyme, diastase (today called amylase), in 1833 by Anselme Payen, while others considered Eduard Buchner's first demonstration of a complex biochemical process alcoholic fermentation in cell-free extracts in 1897 to be the birth of biochemistry.

Gerty Cori and Carl Cori jointly won the Nobel Prize in 1947 for their discovery of the Cori cycle at RPMI.

Some might also point as its beginning to the influential 1842 work by Justus von Liebig, Animal chemistry, or, Organic chemistry in its applications to physiology and pathology, which presented a chemical theory of metabolism, or even earlier to the 18th century studies on fermentation and respiration by Antoine Lavoisier. Many other pioneers in the field who helped to uncover the layers of complexity of biochemistry have been proclaimed founders of modern biochemistry, for example Emil Fischer for his work on the chemistry of proteins, and F. Gowland Hopkins on enzymes and the dynamic nature of biochemistry.

The term "biochemistry" itself is derived from a combination of biology and chemistry. In 1877, Felix Hoppe-Seyler used the term (biochemie in German) as a synonym for physiological chemistry in the foreword to the first issue of Zeitschrift für Physiologische Chemie (Journal of Physiological Chemistry) where he argued for the setting up of institutes dedicated to this field of study. The German chemist Carl Neuberg however is often cited to have been coined the word in 1903, while some credited it to Franz Hofmeister.

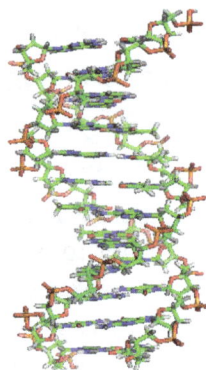

DNA structure (1D65)

It was once generally believed that life and its materials had some essential property or substance (often referred to as the "vital principle") distinct from any found in non-living matter, and it was thought that only living beings could produce the molecules of life. Then, in 1828, Friedrich Wöhler published a paper on the synthesis of urea, proving that organic compounds can be created artificially. Since then, biochemistry has advanced, especially since the mid-20th century, with the development of new techniques such as chromatography, X-ray diffraction, dual polarisation interferometry, NMR spectroscopy, radioisotopic labeling, electron microscopy, and molecular dynamics simulations. These techniques allowed for the discovery and detailed analysis of many molecules and metabolic pathways of the cell, such as glycolysis and the Krebs cycle (citric acid cycle).

Another significant historic event in biochemistry is the discovery of the gene and its role in the transfer of information in the cell. This part of biochemistry is often called molecular biology. In the 1950s, James D. Watson, Francis Crick, Rosalind Franklin, and Maurice Wilkins were instrumental in solving DNA structure and suggesting its relationship with genetic transfer of information. In 1958, George Beadle and Edward Tatum received the Nobel Prize for work in fungi showing that one gene produces one enzyme. In 1988, Colin Pitchfork was the first person convicted of murder with DNA evidence, which led to growth of forensic science. More recently, Andrew Z. Fire and Craig C. Mello received the 2006 Nobel Prize for discovering the role of RNA interference (RNAi), in the silencing of gene expression.

Starting Materials: the Chemical Elements of Life

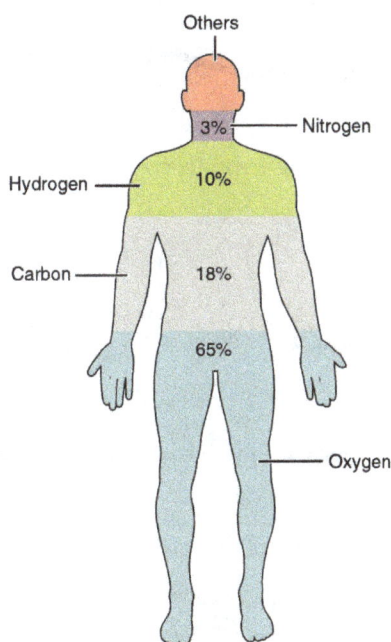

The main elements that compose the human body are shown from most abundant (by mass) to least abundant.

Around two dozen of the 92 naturally occurring chemical elements are essential to various kinds of biological life. Most rare elements on Earth are not needed by life (exceptions being selenium and iodine), while a few common ones (aluminum and titanium) are not used. Most organisms share element needs, but there are a few differences between plants and animals. For example, ocean algae use bromine, but land plants and animals seem to need none. All animals require sodium, but some plants do not. Plants need boron and silicon, but animals may not (or may need ultra-small amounts).

Just six elements—carbon, hydrogen, nitrogen, oxygen, calcium, and phosphorus—make up almost 99% of the mass of living cells, including those in the human body. In addition to the six major elements that compose most of the human body, humans require smaller amounts of possibly 18 more.

Biomolecules

The four main classes of molecules in biochemistry (often called biomolecules) are carbohydrates, lipids, proteins, and nucleic acids. Many biological molecules are polymers: in this terminology, monomers are relatively small micromolecules that are linked together to create large macromolecules known as polymers. When monomers are linked together to synthesize a biological polymer, they undergo a process called dehydration synthesis. Different macromolecules can assemble in larger complexes, often needed for biological activity.

Carbohydrates

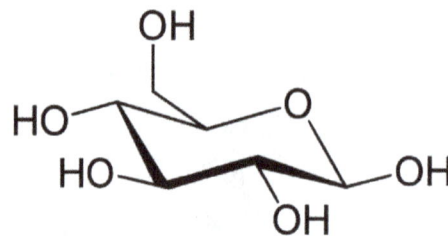

Glucose, a monosaccharide

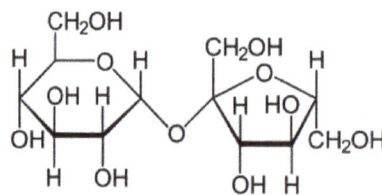

A molecule of sucrose (glucose + fructose), a disaccharide

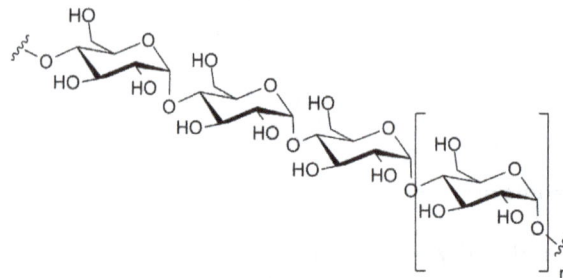

Amylose, a polysaccharide made up of several thousand glucose units

The function of carbohydrates includes energy storage and providing structure. Sugars are carbohydrates, but not all carbohydrates are sugars. There are more carbohydrates on Earth than any other known type of biomolecule; they are used to store energy and genetic information, as well as play important roles in cell to cell interactions and communications.

The simplest type of carbohydrate is a monosaccharide, which among other properties contains carbon, hydrogen, and oxygen, mostly in a ratio of 1:2:1 (generalized formula $C_nH_{2n}O_n$, where n is at least 3). Glucose ($C_6H_{12}O_6$) is one of the most important carbohydrates, others include fructose ($C_6H_{12}O_6$), the sugar commonly associated with the sweet taste of fruits,[a] and deoxyribose ($C_5H_{10}O_4$).

A monosaccharide can switch from the acyclic (open-chain) form to a cyclic form, through a nucleophilic addition reaction between the carbonyl group and one of the hydroxyls of the same molecule. The reaction creates a ring of carbon atoms closed by one bridging oxygen atom. The resulting molecule has an hemiacetal or hemiketal group, depending on whether the linear form was an aldose or a ketose. The reaction is easily reversed, yielding the original open-chain form.

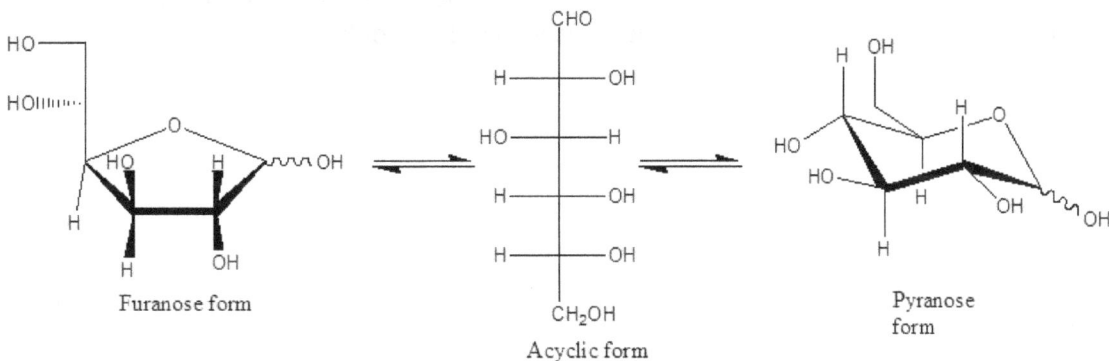

Conversion between the furanose, acyclic, and pyranose forms of D-glucose.

In these cyclic forms, the ring usually has 5 or 6 atoms. These forms are called furanoses and pyranoses, respectively — by analogy with furan and pyran, the simplest compounds with the same carbon-oxygen ring (although they lack the double bonds of these two molecules). For example, the aldohexose glucose may form a hemiacetal linkage between the hydroxyl on carbon 1 and the oxygen on carbon 4, yielding a molecule with a 5-membered ring, called glucofuranose. The same reaction can take place between carbons 1 and 5 to form a molecule with a 6-membered ring, called glucopyranose. Cyclic forms with a 7-atom ring (the same of oxepane), rarely encountered, are called heptoses.

When two monosaccharides undergo dehydration synthesis whereby a molecule of water is released, as two hydrogen atoms and one oxygen atom are lost from the two monosaccharides. The new molecule, consisting of two monosaccharides, is called a disaccharide and is conjoined together by a glycosidic or ether bond. The reverse reaction can also occur, using a molecule of water to split up a disaccharide and break the glycosidic bond; this is termed hydrolysis. The most well-known disaccharide is sucrose, ordinary sugar (in scientific contexts, called table sugar or cane sugar to differentiate it from other sugars). Sucrose consists of a glucose molecule and a fructose molecule joined together. Another important disaccharide is lactose, consisting of a glucose molecule and a galactose molecule. As most humans age, the production of lactase, the enzyme that hydrolyzes lactose back into glucose and galactose, typically decreases. This results in lactase deficiency, also called lactose intolerance.

When a few (around three to six) monosaccharides are joined, it is called an oligosaccharide (oligo-meaning "few"). These molecules tend to be used as markers and signals, as well as having some other uses. Many monosaccharides joined together make a polysaccharide. They can be joined together in one long linear chain, or they may be branched. Two of the most common polysaccharides are cellulose and glycogen, both consisting of repeating glucose monomers. Examples are Cellulose which is an important structural component of plant's cell walls, and glycogen, used as a form of energy storage in animals.

Sugar can be characterized by having reducing or non-reducing ends. A reducing end of a carbohydrate is a carbon atom that can be in equilibrium with the open-chain aldehyde (aldose) or keto form (ketose). If the joining of monomers takes place at such a carbon atom, the free hydroxy group of the pyranose or furanose form is exchanged with an OH-side-chain of another sugar, yielding a full acetal. This prevents opening of the chain to the aldehyde or keto form and renders the modified residue non-reducing. Lactose contains a reducing end at its glucose moiety, whereas the galactose moiety form a full acetal with the C4-OH group of glucose. Saccharose does not have a reducing end because of full acetal formation between the aldehyde carbon of glucose (C1) and the keto carbon of fructose (C2).

Lipids

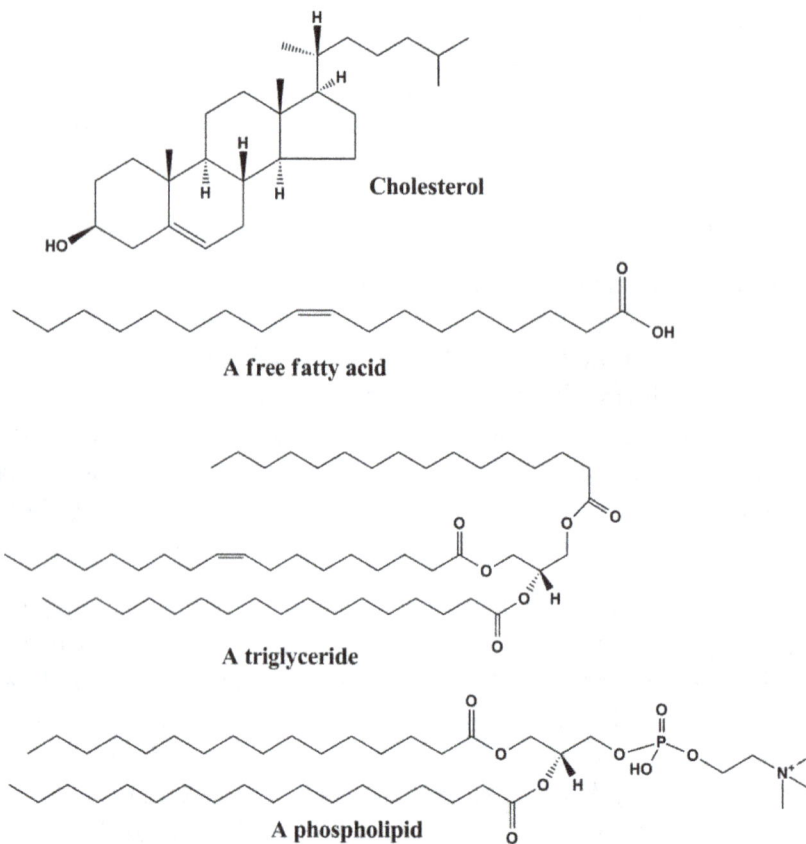

Structures of some common lipids. At the top are cholesterol and oleic acid. The middle structure is a triglyceride composed of oleoyl, stearoyl, and palmitoyl chains attached to a glycerol backbone. At the bottom is the common phospholipid, phosphatidylcholine.

Lipids comprises a diverse range of molecules and to some extent is a catchall for relatively water-insoluble or nonpolar compounds of biological origin, including waxes, fatty acids, fatty-acid derived phospholipids, sphingolipids, glycolipids, and terpenoids (e.g., retinoids and steroids). Some lipids are linear aliphatic molecules, while others have ring structures. Some are aromatic, while others are not. Some are flexible, while others are rigid.

Lipids are usually made from one molecule of glycerol combined with other molecules. In triglycerides, the main group of bulk lipids, there is one molecule of glycerol and three fatty acids. Fatty acids are considered the monomer in that case, and may be saturated (no double bonds in the carbon chain) or unsaturated (one or more double bonds in the carbon chain).

Most lipids have some polar character in addition to being largely nonpolar. In general, the bulk of their structure is nonpolar or hydrophobic ("water-fearing"), meaning that it does not interact well with polar solvents like water. Another part of their structure is polar or hydrophilic ("water-loving") and will tend to associate with polar solvents like water. This makes them amphiphilic molecules (having both hydrophobic and hydrophilic portions). In the case of cholesterol, the polar group is a mere -OH (hydroxyl or alcohol). In the case of phospholipids, the polar groups are considerably larger and more polar, as described below.

Lipids are an integral part of our daily diet. Most oils and milk products that we use for cooking and eating like butter, cheese, ghee etc., are composed of fats. Vegetable oils are rich in various polyunsaturated fatty acids (PUFA). Lipid-containing foods undergo digestion within the body and are broken into fatty acids and glycerol, which are the final degradation products of fats and lipids. Lipids, especially phospholipids, are also used in various pharmaceutical products, either as co-solubilisers (e.g., in parenteral infusions) or else as drug carrier components (e.g., in a liposome or transfersome).

Proteins

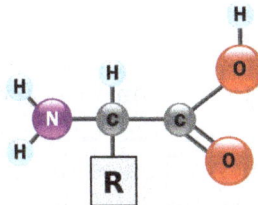

The general structure of an α-amino acid, with the amino group on the left and the carboxyl group on the right.

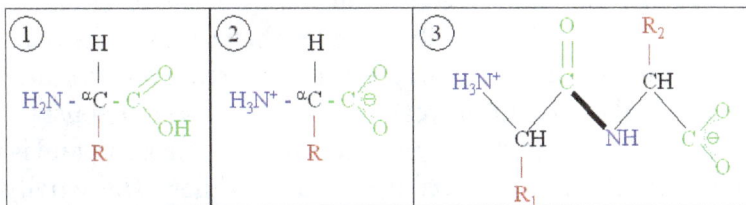

Generic amino acids (1) in neutral form, (2) as they exist physiologically, and (3) joined together as a dipeptide.

Proteins are very large molecules – macro-biopolymers – made from monomers called amino acids. An amino acid consists of a carbon atom bound to four groups. One is an amino group, $-NH_2$, and one is a carboxylic acid group, —COOH (although these exist as —NH$_3^+$ and —COO$^-$ under physiologic conditions). The third is a simple hydrogen atom. The fourth is commonly denoted "—R" and is different for each amino acid. There are 20 standard amino acids, each containing a carboxyl group, an amino group, and a side-chain (known as an "R" group). The "R" group is what makes each amino acid different, and the properties of the side-chains greatly influence the overall three-dimensional conformation of a protein. Some amino acids have functions by themselves or in a modified form; for instance, glutamate functions as an important neurotransmitter. Amino acids can be joined via a peptide bond. In this dehydration synthesis, a water molecule is removed and the peptide bond connects the nitrogen of one amino acid's amino group to the carbon of the other's carboxylic acid group. The resulting molecule is called a dipeptide, and short stretches of amino acids (usually, fewer than thirty) are called peptides or polypeptides. Longer stretches merit the title proteins. As an example, the important blood serum protein albumin contains 585 amino acid residues.

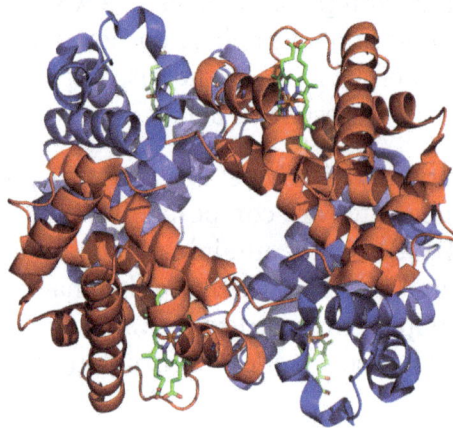

A schematic of hemoglobin. The red and blue ribbons represent the protein globin; the green structures are the heme groups.

Some proteins perform largely structural roles. For instance, movements of the proteins actin and myosin ultimately are responsible for the contraction of skeletal muscle. One property many proteins have is that they specifically bind to a certain molecule or class of molecules—they may be extremely selective in what they bind. Antibodies are an example of proteins that attach to one specific type of molecule. In fact, the enzyme-linked immunosorbent assay (ELISA), which uses antibodies, is one of the most sensitive tests modern medicine uses to detect various biomolecules. Probably the most important proteins, however, are the enzymes. Virtually every reaction in a living cell requires an enzyme to lower the activation energy of the reaction. These molecules recognize specific reactant molecules called substrates; they then catalyze the reaction between them. By lowering the activation energy, the enzyme speeds up that reaction by a rate of 10 or more; a reaction that would normally take over 3,000 years to complete spontaneously might take less than a second with an enzyme. The enzyme itself is not used up in the process, and is free to catalyze the same reaction with a new set of substrates. Using various modifiers, the activity of the enzyme can be regulated, enabling control of the biochemistry of the cell as a whole.

The structure of proteins is traditionally described in a hierarchy of four levels. The primary structure of a protein simply consists of its linear sequence of amino acids; for instance, "alanine-glycine-tryptophan-serine-glutamate-asparagine-glycine-lysine-...". Secondary structure is concerned with local morphology (morphology being the study of structure). Some combinations of amino acids will tend to curl up in a coil called an α-helix or into a sheet called a β-sheet; some α-helixes can be seen in the hemoglobin schematic above. Tertiary structure is the entire three-dimensional shape of the protein. This shape is determined by the sequence of amino acids. In fact, a single change can change the entire structure. The alpha chain of hemoglobin contains 146 amino acid residues; substitution of the glutamate residue at position 6 with a valine residue changes the behavior of hemoglobin so much that it results in sickle-cell disease. Finally, quaternary structure is concerned with the structure of a protein with multiple peptide subunits, like hemoglobin with its four subunits. Not all proteins have more than one subunit.

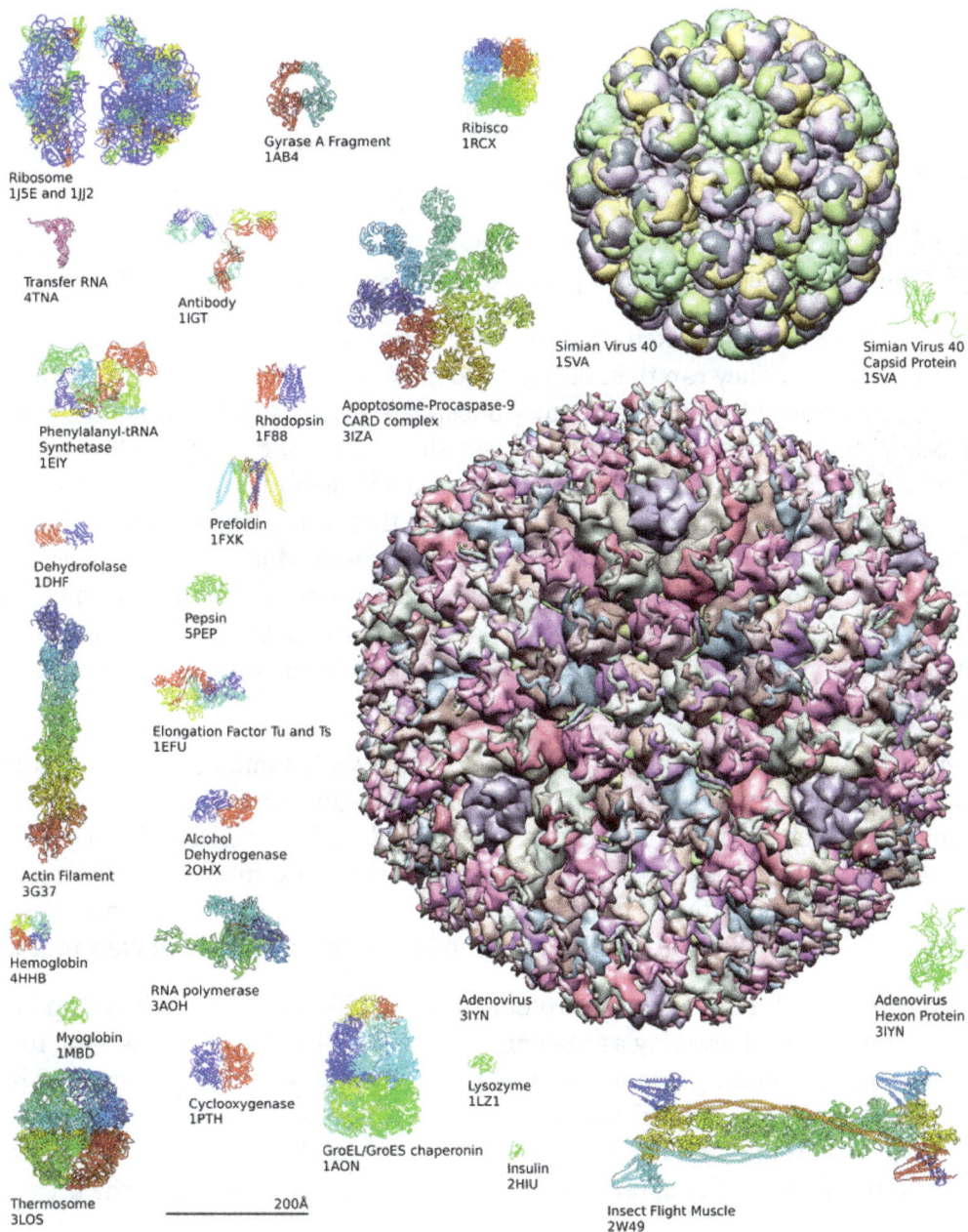

Examples of protein structures from the Protein Data Bank

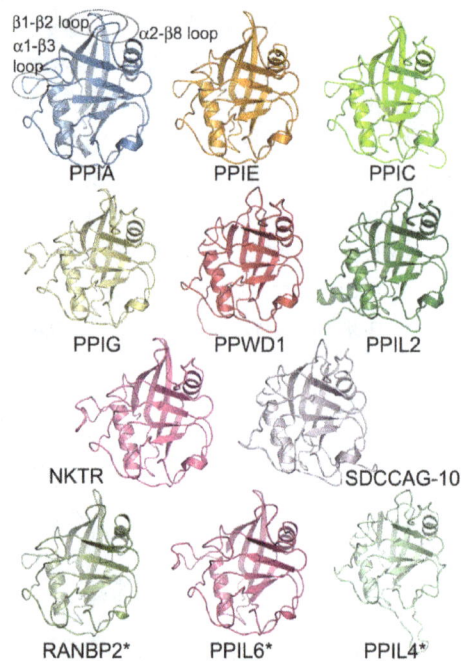

β1-β2 loop
α1-β3 loop
α2-β8 loop

PPIA   PPIE   PPIC

PPIG   PPWD1   PPIL2

NKTR   SDCCAG-10

RANBP2*   PPIL6*   PPIL4*

Members of a protein family, as represented by the structures of the isomerase domains.

Ingested proteins are usually broken up into single amino acids or dipeptides in the small intestine, and then absorbed. They can then be joined to make new proteins. Intermediate products of glycolysis, the citric acid cycle, and the pentose phosphate pathway can be used to make all twenty amino acids, and most bacteria and plants possess all the necessary enzymes to synthesize them. Humans and other mammals, however, can synthesize only half of them. They cannot synthesize isoleucine, leucine, lysine, methionine, phenylalanine, threonine, tryptophan, and valine. These are the essential amino acids, since it is essential to ingest them. Mammals do possess the enzymes to synthesize alanine, asparagine, aspartate, cysteine, glutamate, glutamine, glycine, proline, serine, and tyrosine, the nonessential amino acids. While they can synthesize arginine and histidine, they cannot produce it in sufficient amounts for young, growing animals, and so these are often considered essential amino acids.

If the amino group is removed from an amino acid, it leaves behind a carbon skeleton called an α-keto acid. Enzymes called transaminases can easily transfer the amino group from one amino acid (making it an α-keto acid) to another α-keto acid (making it an amino acid). This is important in the biosynthesis of amino acids, as for many of the pathways, intermediates from other biochemical pathways are converted to the α-keto acid skeleton, and then an amino group is added, often via transamination. The amino acids may then be linked together to make a protein.

A similar process is used to break down proteins. It is first hydrolyzed into its component amino acids. Free ammonia (NH_3), existing as the ammonium ion (NH_4^+) in blood, is toxic to life forms. A suitable method for excreting it must therefore exist. Different tactics have evolved in different animals, depending on the animals' needs. Unicellular organisms, of course, simply release the ammonia into the environment. Likewise, bony fish can release the ammonia into the water where it is quickly diluted. In general, mammals convert the ammonia into urea, via the urea cycle.

In order to determine whether two proteins are related, or in other words to decide whether they are homologous or not, scientists use sequence-comparison methods. Methods like sequence alignments and structural alignments are powerful tools that help scientists identify homologies between related molecules. The relevance of finding homologies among proteins goes beyond forming an evolutionary pattern of protein families. By finding how similar two protein sequences are, we acquire knowledge about their structure and therefore their function.

Nucleic Acids

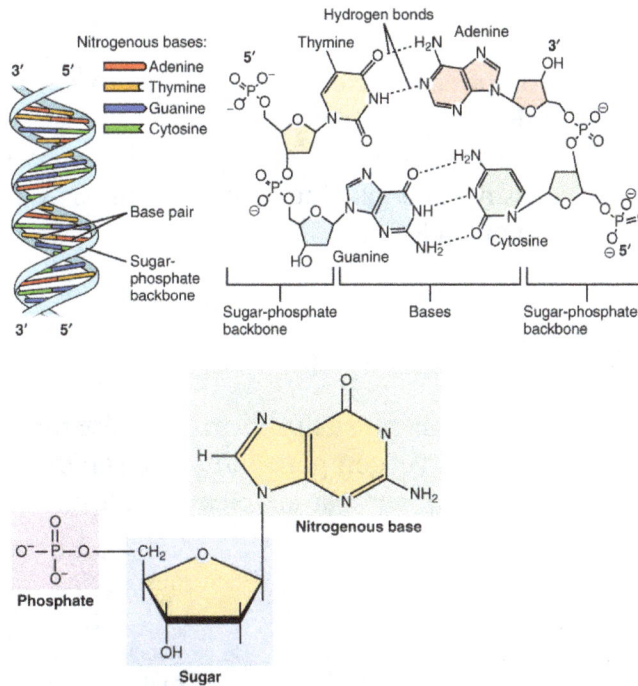

The structure of deoxyribonucleic acid (DNA), the picture shows the monomers being put together.

Nucleic acids, so called because of its prevalence in cellular nuclei, is the generic name of the family of biopolymers. They are complex, high-molecular-weight biochemical macromolecules that can convey genetic information in all living cells and viruses. The monomers are called nucleotides, and each consists of three components: a nitrogenous heterocyclic base (either a purine or a pyrimidine), a pentose sugar, and a phosphate group.

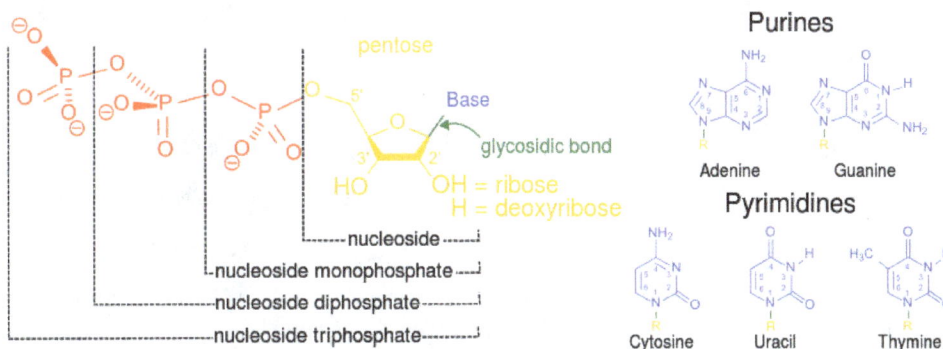

Structural elements of common nucleic acid constituents. Because they contain at least one phosphate group, the compounds marked nucleoside monophosphate, nucleoside diphosphate and nucleoside triphosphate are all nucleotides (not simply phosphate-lacking nucleosides).

The most common nucleic acids are deoxyribonucleic acid (DNA) and ribonucleic acid (RNA). The phosphate group and the sugar of each nucleotide bond with each other to form the backbone of the nucleic acid, while the sequence of nitrogenous bases stores the information. The most common nitrogenous bases are adenine, cytosine, guanine, thymine, and uracil. The nitrogenous bases of each strand of a nucleic acid will form hydrogen bonds with certain other nitrogenous bases in a complementary strand of nucleic acid (similar to a zipper). Adenine binds with thymine and uracil; Thymine binds only with adenine; and cytosine and guanine can bind only with one another.

Aside from the genetic material of the cell, nucleic acids often play a role as second messengers, as well as forming the base molecule for adenosine triphosphate (ATP), the primary energy-carrier molecule found in all living organisms. Also, the nitrogenous bases possible in the two nucleic acids are different: adenine, cytosine, and guanine occur in both RNA and DNA, while thymine occurs only in DNA and uracil occurs in RNA.

Metabolism

Carbohydrates as Energy Source

Glucose is the major energy source in most life forms. For instance, polysaccharides are broken down into their monomers (glycogen phosphorylase removes glucose residues from glycogen). Disaccharides like lactose or sucrose are cleaved into their two component monosaccharides.

Glycolysis (Anaerobic)

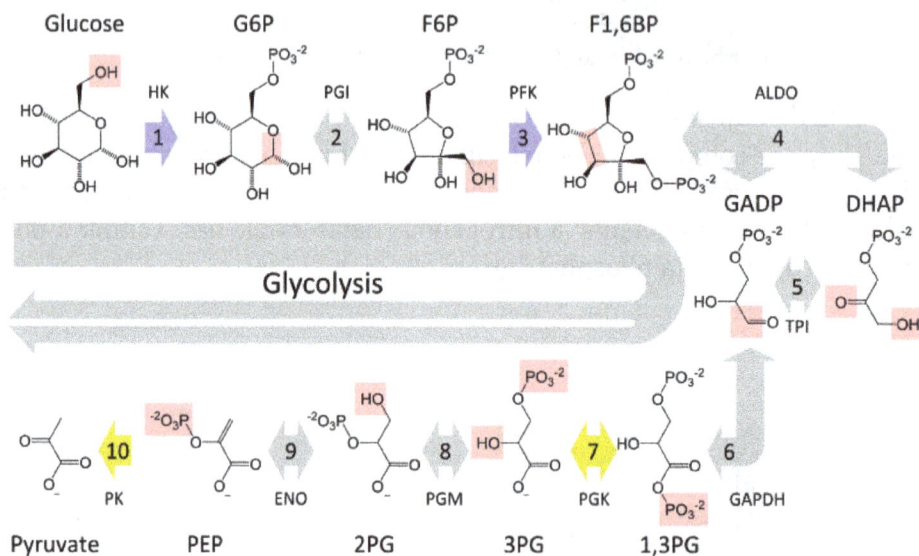

The metabolic pathway of glycolysis converts glucose to pyruvate by via a series of intermediate metabolites. Each chemical modification (red box) is performed by a different enzyme. Steps 1 and 3 consume ATP (blue) and steps 7 and 10 produce ATP (yellow). Since steps 6-10 occur twice per glucose molecule, this leads to a net production of ATP.

Glucose is mainly metabolized by a very important ten-step pathway called glycolysis, the net result of which is to break down one molecule of glucose into two molecules of pyruvate. This also produces a net two molecules of ATP, the energy currency of cells, along with two reducing equivalents of converting NAD^+ (nicotinamide adenine dinucleotide:oxidised form) to NADH (nicotinamide adenine dinucleotide:reduced form). This does not require oxygen; if no oxygen is available (or the cell cannot use oxygen), the NAD is restored by converting the pyruvate to lactate (lactic acid) (e.g., in humans) or to ethanol plus carbon dioxide (e.g., in yeast). Other monosaccharides like galactose and fructose can be converted into intermediates of the glycolytic pathway.

Aerobic

In aerobic cells with sufficient oxygen, as in most human cells, the pyruvate is further metabolized. It is irreversibly converted to acetyl-CoA, giving off one carbon atom as the waste product carbon dioxide, generating another reducing equivalent as NADH. The two molecules acetyl-CoA (from one molecule of glucose) then enter the citric acid cycle, producing two more molecules of ATP, six more NADH molecules and two reduced (ubi)quinones (via $FADH_2$ as enzyme-bound cofactor), and releasing the remaining carbon atoms as carbon dioxide. The produced NADH and quinol molecules then feed into the enzyme complexes of the respiratory chain, an electron transport system transferring the electrons ultimately to oxygen and conserving the released energy in the form of a proton gradient over a membrane (inner mitochondrial membrane in eukaryotes). Thus, oxygen is reduced to water and the original electron acceptors NAD^+ and quinone are regenerated. This is why humans breathe in oxygen and breathe out carbon dioxide. The energy released from transferring the electrons from high-energy states in NADH and quinol is conserved first as proton gradient and converted to ATP via ATP synthase. This generates an additional 28 molecules of ATP (24 from the 8 NADH + 4 from the 2 quinols), totaling to 32 molecules of ATP conserved per degraded glucose (two from glycolysis + two from the citrate cycle). It is clear that using oxygen to completely oxidize glucose provides an organism with far more energy than any oxygen-independent metabolic feature, and this is thought to be the reason why complex life appeared only after Earth's atmosphere accumulated large amounts of oxygen.

Gluconeogenesis

In vertebrates, vigorously contracting skeletal muscles (during weightlifting or sprinting, for example) do not receive enough oxygen to meet the energy demand, and so they shift to anaerobic metabolism, converting glucose to lactate. The liver regenerates the glucose, using a process called gluconeogenesis. This process is not quite the opposite of glycolysis, and actually requires three times the amount of energy gained from glycolysis (six molecules of ATP are used, compared to the two gained in glycolysis). Analogous to the above reactions, the glucose produced can then undergo glycolysis in tissues that need energy, be stored as glycogen (or starch in plants), or be converted to other monosaccharides or joined into di- or oligosaccharides. The combined pathways of glycolysis during exercise, lactate's crossing via the bloodstream to the liver, subsequent gluconeogenesis and release of glucose into the bloodstream is called the Cori cycle.

Relationship to Other "Molecular-Scale" Biological Sciences

Researchers in biochemistry use specific techniques native to biochemistry, but increasingly combine these with techniques and ideas developed in the fields of genetics, molecular biology and biophysics. There has never been a hard-line among these disciplines in terms of content and technique. Today, the terms molecular biology and biochemistry are nearly interchangeable. The following figure is a schematic that depicts one possible view of the relationship between the fields:

- Biochemistry is the study of the chemical substances and vital processes occurring in living organisms. Biochemists focus heavily on the role, function, and structure of biomolecules. The study of the chemistry behind biological processes and the synthesis of biologically active molecules are examples of biochemistry.

- Genetics is the study of the effect of genetic differences on organisms. Often this can be inferred by the absence of a normal component (e.g., one gene). The study of "mutants" – organisms with a changed gene that leads to the organism being different with respect to the so-called "wild type" or normal phenotype. Genetic interactions (epistasis) can often confound simple interpretations of such "knock-out" or "knock-in" studies.

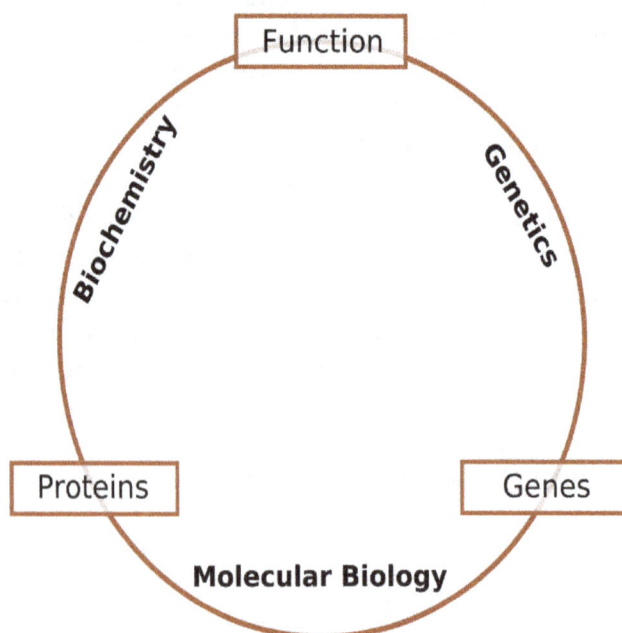

Schematic relationship between biochemistry, genetics, and molecular biology

- Molecular biology is the study of molecular underpinnings of the process of replication, transcription and translation of the genetic material. The central dogma of molecular biology where genetic material is transcribed into RNA and then translated into protein, despite being an oversimplified picture of molecular biology, still provides a good starting point for understanding the field. This picture, however, is undergoing revision in light of emerging novel roles for RNA.

- Chemical biology seeks to develop new tools based on small molecules that allow minimal perturbation of biological systems while providing detailed information about their function. Further, chemical biology employs biological systems to create non-natural hybrids between biomolecules and synthetic devices (for example emptied viral capsids that can deliver gene therapy or drug molecules).

Essential Elements of Biochemistry

This chapter will deal in the major and minor categories of biological macromolecules. The major macromolecules discussed here are carbohydrates, nucleic acids, proteins and lipids. Many forms and functions of lipids are also discussed in the chapter like lipid microdomain, lipidomics, phenolic lipids and protein-lipid interactions. The chapter is written in a comprehensive manner to provide easy and simplistic explanations of concepts.

Carbohydrate

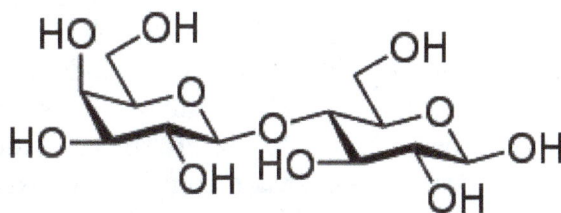

Lactose is a disaccharide found in milk. It consists of a molecule of D-galactose and a molecule of D-glucosebonded by beta-1-4 glycosidic linkage. It has a formula of C12H22O11.

A carbohydrate is a biological molecule consisting of carbon (C), hydrogen (H) and oxygen (O) atoms, usually with a hydrogen–oxygen atom ratio of 2:1 (as in water); in other words, with the empirical formula $C_m(H_2O)_n$ (where m could be different from n). Some exceptions exist; for example, deoxyribose, a sugar component of DNA, has the empirical formula $C_5H_{10}O_4$. Carbohydrates are technically hydrates of carbon; structurally it is more accurate to view them as polyhydroxy aldehydes and ketones.

The term is most common in biochemistry, where it is a synonym of saccharide, a group that includes sugars, starch, and cellulose. The saccharides are divided into four chemical groups: monosaccharides, disaccharides, oligosaccharides, and polysaccharides. In general, the monosaccharides and disaccharides, which are smaller (lower molecular weight) carbohydrates, are commonly referred to as sugars. The word saccharide comes from the Greek word meaning "sugar." While the scientific nomenclature of carbohydrates is complex, the names of the monosaccharides and disaccharides very often end in the suffix -ose. For example, grape sugar is the monosaccharide glucose, cane sugar is the disaccharide sucrose, and milk sugar is the disaccharide lactose.

Carbohydrates perform numerous roles in living organisms. Polysaccharides serve for the storage of energy (e.g. starch and glycogen) and as structural components (e.g. cellulose in plants and chitin in arthropods). The 5-carbon monosaccharide ribose is an important component of coenzymes (e.g. ATP, FAD and NAD) and the backbone of the genetic molecule known as RNA. The related deoxyribose is a component of DNA. Saccharides and their derivatives include many other important biomolecules that play key roles in the immune system, fertilization, preventing pathogenesis, blood clotting, and development.

In food science and in many informal contexts, the term carbohydrate often means any food that is particularly rich in the complex carbohydrate starch (such as cereals, bread and pasta) or simple carbohydrates, such as sugar (found in candy, jams, and desserts).

Often in lists of nutritional information, such as the USDA National Nutrient Database, the term "carbohydrate" (or "carbohydrate by difference") is used for everything other than water, protein, fat, ash, and ethanol. This will include chemical compounds such as acetic or lactic acid, which are not normally considered carbohydrates. It also includes "dietary fiber" which is a carbohydrate but which does not contribute much in the way of food energy (calories), even though it is often included in the calculation of total food energy just as though it were a sugar.

Structure

Formerly the name "carbohydrate" was used in chemistry for any compound with the formula C_m $(H_2O)_n$. Following this definition, some chemists considered formaldehyde (CH_2O) to be the simplest carbohydrate, while others claimed that title for glycolaldehyde. Today, the term is generally understood in the biochemistry sense, which excludes compounds with only one or two carbons and includes many biological carbohydrates which deviate from this formula. For example, while the above representative formulas would seem to capture the commonly known carbohydrates, ubiquitous and abundant carbohydrates often deviate from this. For example, carbohydrates often display chemical groups such as: N-acetyl (e.g. chitin), sulphate (e.g. glycosaminoglycans), carboxylic acid (e.g. sialic acid) and deoxy modifications (e.g. fucose and sialic acid).

Natural saccharides are generally built of simple carbohydrates called monosaccharides with general formula $(CH_2O)_n$ where n is three or more. A typical monosaccharide has the structure H–$(CHOH)_x(C=O)–(CHOH)_y$–H, that is, an aldehyde or ketone with many hydroxyl groups added, usually one on each carbon atom that is not part of the aldehyde or ketone functional group. Examples of monosaccharides are glucose, fructose, and glyceraldehydes. However, some biological substances commonly called "monosaccharides" do not conform to this formula (e.g. uronic acids and deoxy-sugars such as fucose) and there are many chemicals that do conform to this formula but are not considered to be monosaccharides (e.g. formaldehyde CH_2O and inositol $(CH_2O)_6$).

The open-chain form of a monosaccharide often coexists with a closed ring form where the aldehyde/ketone carbonyl group carbon (C=O) and hydroxyl group (–OH) react forming a hemiacetal with a new C–O–C bridge.

Monosaccharides can be linked together into what are called polysaccharides (or oligosaccharides) in a large variety of ways. Many carbohydrates contain one or more modified monosaccharide units that have had one or more groups replaced or removed. For example, deoxyribose, a component of DNA, is a modified version of ribose; chitin is composed of repeating units of N-acetyl glucosamine, a nitrogen-containing form of glucose.

Division

Carbohydrates are polyhydroxy aldehydes, ketones, alcohols, acids, their simple derivatives and their polymers having linkages of the acetal type. They may be classified according to their degree of polymerization and may be divided initially into three principal groups, namely sugars, oligo-saccharides and polysaccharides

The major dietary carbohydrates		
Class (DP*)	**Subgroup**	**Components**
Sugars (1–2)	Monosaccharides	Glucose, galactose, fructose, xylose
	Disaccharides	Sucrose, lactose, maltose, trehalose
	Polyols	Sorbitol, mannitol
Oligosaccharides (3–9)	Malto-oligosaccharides	Maltodextrins
	Other oligosaccharides	Raffinose, stachyose, fructo-oligosaccharides
Polysaccharides (>9)	Starch	Amylose, amylopectin, modified starches
	Non-starch polysaccharides	Cellulose, hemicellulose, pectins, hydrocolloids

DP * = Degree of polymerization

Monosaccharides

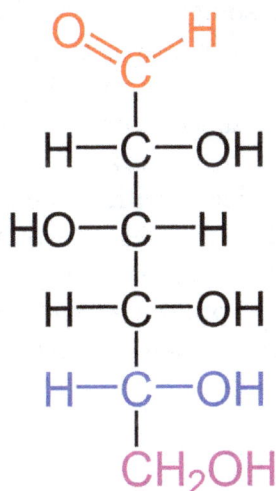

D-glucose is an aldohexose with the formula $(C \cdot H_2O)_6$. The red atoms highlight the aldehyde group and the blue atoms highlight the asymmetric center furthest from the aldehyde; because this -OH is on the right of the Fischer projection, this is a D sugar.

Monosaccharides are the simplest carbohydrates in that they cannot be hydrolyzed to smaller carbohydrates. They are aldehydes or ketones with two or more hydroxyl groups. The general chemical formula of an unmodified monosaccharide is $(C \cdot H_2O)_n$, literally a "carbon hydrate." Monosaccharides are important fuel molecules as well as building blocks for nucleic acids. The smallest monosaccharides, for which n=3, are dihydroxyacetone and D- and L-glyceral-dehydes.

Classification of Monosaccharides

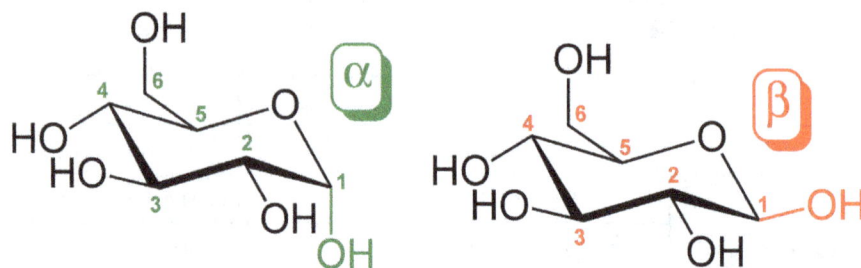

The α and β anomers of glucose. Note the position of the hydroxyl group (red or green) on the anomeric carbon relative to the CH$_2$OH group bound to carbon 5: they either have identical absolute configurations (R,R or S,S) (α), or opposite absolute configurations (R,S or S,R) (β).

Monosaccharides are classified according to three different characteristics: the placement of its carbonyl group, the number of carbon atoms it contains, and its chiral handedness. If the carbonyl group is an aldehyde, the monosaccharide is an aldose; if the carbonyl group is a ketone, the monosaccharide is a ketose. Monosaccharides with three carbon atoms are called trioses, those with four are called tetroses, five are called pentoses, six are hexoses, and so on. These two systems of classification are often combined. For example, glucose is an aldohexose (a six-carbon aldehyde), ribose is an aldopentose (a five-carbon aldehyde), and fructose is a ketohexose (a six-carbon ketone).

Each carbon atom bearing a hydroxyl group (-OH), with the exception of the first and last carbons, are asymmetric, making them stereo centers with two possible configurations each (R or S). Because of this asymmetry, a number of isomers may exist for any given monosaccharide formula. Using Le Bel-van't Hoff rule, the aldohexose D-glucose, for example, has the formula (C·H$_2$O)$_6$, of which four of its six carbons atoms are stereogenic, making D-glucose one of 2=16 possible stereoisomers. In the case of glyceraldehydes, an aldotriose, there is one pair of possible stereoisomers, which are enantiomers and epimers. 1, 3-dihydroxyacetone, the ketose corresponding to the aldose glyceraldehydes, is a symmetric molecule with no stereo centers. The assignment of D or L is made according to the orientation of the asymmetric carbon furthest from the carbonyl group: in a standard Fischer projection if the hydroxyl group is on the right the molecule is a D sugar, otherwise it is an L sugar. The "D-" and "L-" prefixes should not be confused with "d-" or "l-", which indicate the direction that the sugar rotates plane polarized light. This usage of "d-" and "l-" is no longer followed in carbohydrate chemistry.

Ring-straight Chain Isomerism

The aldehyde or ketone group of a straight-chain monosaccharide will react reversibly with a hydroxyl group on a different carbon atom to form a hemiacetal or hemiketal, forming a heterocyclic ring with an oxygen bridge between two carbon atoms. Rings with five and six atoms are called furanose and pyranose forms, respectively, and exist in equilibrium with the straight-chain form.

During the conversion from straight-chain form to the cyclic form, the carbon atom containing the carbonyl oxygen, called the anomeric carbon, becomes a stereogenic center with two possible configurations: The oxygen atom may take a position either above or below the plane of the ring. The resulting possible pair of stereoisomers is called anomers. In the α anomer, the -OH substituent on the anomeric carbon rests on the opposite side (trans) of the ring from the CH_2OH side branch. The alternative form, in which the CH_2OH substituent and the anomeric hydroxyl are on the same side (cis) of the plane of the ring, is called the β anomer.

Use in Living Organisms

Monosaccharides are the major source of fuel for metabolism, being used both as an energy source (glucose being the most important in nature) and in biosynthesis. When monosaccharides are not immediately needed by many cells they are often converted to more space-efficient forms, often polysaccharides. In many animals, including humans, this storage form is glycogen, especially in liver and muscle cells. In plants, starch is used for the same purpose. The most abundant carbohydrate, cellulose, is a structural component of the cell wall of plants and many forms of algae. Ribose is a component of RNA. Deoxyribose is a component of DNA. Lyxose is a component of lyxoflavin found in the human heart. Ribulose and xylulose occur in the pentose phosphate pathway. Galactose, a component of milk sugar lactose, is found in galactolipids in plant cell membranes and in glycoproteins in many tissues. Mannose occurs in human metabolism, especially in the glycosylation of certain proteins. Fructose, or fruit sugar, is found in many plants and in humans, it is metabolized in the liver, absorbed directly into the intestines during digestion, and found in semen. Trehalose, a major sugar of insects, is rapidly hydrolyzed into two glucose molecules to support continuous flight.

Disaccharides

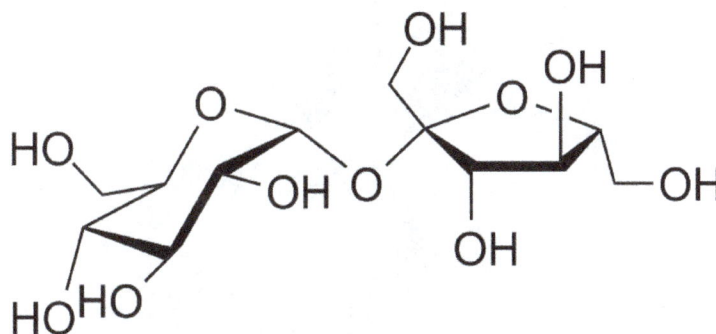

Sucrose, also known as table sugar, is a common disaccharide. It is composed of two monosaccharides: D-glucose (left) and D-fructose (right).

Two joined monosaccharides are called a disaccharide and these are the simplest polysaccharides. Examples include sucrose and lactose. They are composed of two monosaccharide units bound together by a covalent bond known as a glycosidic linkage formed via a dehydration reaction, resulting in the loss of a hydrogen atom from one monosaccharide and a hydroxyl group from the other. The formula of unmodified disaccharides is $C_{12}H_{22}O_{11}$. Although there are numerous kinds of disaccharides, a handful of disaccharides are particularly notable.

Sucrose, pictured to the right, is the most abundant disaccharide, and the main form in which carbohydrates are transported in plants. It is composed of one D-glucose molecule and one D-fructose molecule. The systematic name for sucrose, O-α-D-glucopyranosyl-(1→2)-D-fructofuranoside, indicates four things:

- Its monosaccharides: glucose and fructose

- Their ring types: glucose is a pyranose and fructose is a furanose

- How they are linked together: the oxygen on carbon number 1 (C1) of α-D-glucose is linked to the C2 of D-fructose.

- The -oside suffix indicates that the anomeric carbon of both monosaccharides participates in the glycosidic bond.

Lactose, a disaccharide composed of one D-galactose molecule and one D-glucose molecule, occurs naturally in mammalian milk. The systematic name for lactose is O-β-D-galactopyranosyl-(1→4)-D-glucopyranose. Other notable disaccharides include maltose (two D-glucoses linked α-1,4) and cellulobiose (two D-glucoses linked β-1,4). Disaccharides can be classified into two types: reducing and non-reducing disaccharides. If the functional group is present in bonding with another sugar unit, it is called a reducing disaccharide or biose.

Nutrition

Grain products: rich sources of carbohydrates

Carbohydrate consumed in food yields 3.87 calories of energy per gram for simple sugars, and 3.57 to 4.12 calories per gram for complex carbohydrate in most other foods. Relatively high levels of carbohydrate are associated with processed foods or refined foods made from plants, including sweets, cookies and candy, table sugar, honey, soft drinks, breads and crackers, jams and fruit products, pastas and breakfast cereals. Lower amounts of carbohydrate are usually associated with unrefined foods, including beans, tubers, rice, and unrefined fruit. Animal-based foods generally have the lowest carbohydrate levels, although milk does contain a high proportion of lactose.

Carbohydrates are a common source of energy in living organisms; however, no carbohydrate is an essential nutrient in humans. Humans are able to obtain all of their energy requirement from protein and fats, though the potential for some negative health effects of extreme carbohydrate restriction remains, as the issue has not been studied extensively so far. However, in the case of dietary fiber – indigestible carbohydrates which are not a source of energy – inadequate intake can lead to significant increases in mortality.

Following a diet consisting of very low amounts of daily carbohydrate for several days will usually result in higher levels of blood ketone bodies than an isocaloric diet with similar protein content. This relatively high level of ketone bodies is commonly known as ketosis and is very often confused with the potentially fatal condition often seen in type 1 diabetics known as diabetic ketoacidosis. Somebody suffering ketoacidosis will have much higher levels of blood ketone bodies along with high blood sugar, dehydration and electrolyte imbalance.

Long-chain fatty acids cannot cross the blood–brain barrier, but the liver can break these down to produce ketones. However, the medium-chain fatty acids octanoic and heptanoic acids can cross the barrier and be used by the brain, which normally relies upon glucose for its energy. Gluconeogenesis allows humans to synthesize some glucose from specific amino acids: from the glycerol backbone in triglycerides and in some cases from fatty acids.

Organisms typically cannot metabolize all types of carbohydrate to yield energy. Glucose is a nearly universal and accessible source of energy. Many organisms also have the ability to metabolize other monosaccharides and disaccharides but glucose is often metabolized first. In Escherichia coli, for example, the lac operon will express enzymes for the digestion of lactose when it is present, but if both lactose and glucose are present the lac operon is repressed, resulting in the glucose being used first. Polysaccharides are also common sources of energy. Many organisms can easily break down starches into glucose; most organisms, however, cannot metabolize cellulose or other polysaccharides like chitin and arabinoxylans. These carbohydrate types can be metabolized by some bacteria and protists. Ruminants and termites, for example, use microorganisms to process cellulose. Even though these complex carbohydrates are not very digestible, they represent an important dietary element for humans, called dietary fiber. Fiber enhances digestion, among other benefits.

Based on the effects on risk of heart disease and obesity, the Institute of Medicine recommends that American and Canadian adults get between 45–65% of dietary energy from carbohydrates. The Food and Agriculture Organization and World Health Organization jointly recommend that national dietary guidelines set a goal of 55–75% of total energy from carbohydrates, but only 10% directly from sugars (their term for simple carbohydrates).

Classification

Nutritionists often refer to carbohydrates as either simple or complex. However, the exact distinction between these groups can be ambiguous. The term complex carbohydrate was first used in the U.S. Senate Select Committee on Nutrition and Human Needs publication Dietary Goals for the United States (1977) where it was intended to distinguish sugars from other carbohydrates (which were perceived to be nutritionally superior). However, the report put "fruit, vegetables and wholegrains" in the complex carbohydrate column, despite the fact that these may contain sugars as well as polysaccharides. This confusion persists as today some nutritionists use the term complex carbohydrate to refer to any sort of digestible saccharide present in a whole food, where fiber, vitamins and minerals are also found (as opposed to processed carbohydrates, which provide energy but few other nutrients). The standard usage, however, is to classify carbohydrates chemically: simple if they are sugars (monosaccharides and disaccharides) and complex if they are polysaccharides (or oligosaccharides).

In any case, the simple vs. complex chemical distinction has little value for determining the nutritional quality of carbohydrates. Some simple carbohydrates (e.g. fructose) raise blood glucose slowly, while some complex carbohydrates (starches), especially if processed, raise blood sugar rapidly. The speed of digestion is determined by a variety of factors including which other nutrients are consumed with the carbohydrate, how the food is prepared, individual differences in metabolism, and the chemistry of the carbohydrate.

The USDA's Dietary Guidelines for Americans 2010 call for moderate- to high-carbohydrate consumption from a balanced diet that includes six one-ounce servings of grain foods each day, at least half from whole grain sources and the rest from enriched.

The glycemic index (GI) and glycemic load concepts have been developed to characterize food behavior during human digestion. They rank carbohydrate-rich foods based on the rapidity and magnitude of their effect on blood glucose levels. Glycemic index is a measure of how quickly food glucose is absorbed, while glycemic load is a measure of the total absorbable glucose in foods. The insulin index is a similar, more recent classification method that ranks foods based on their effects on blood insulin levels, which are caused by glucose (or starch) and some amino acids in food.

Metabolism

Carbohydrate metabolism denotes the various biochemical processes responsible for the formation, breakdown and interconversion of carbohydrates in living organisms.

The most important carbohydrate is glucose, a simple sugar (monosaccharide) that is metabolized by nearly all known organisms. Glucose and other carbohydrates are part of a wide variety of metabolic pathways across species: plants synthesize carbohydrates from carbon dioxide and water by photosynthesis storing the absorbed energy internally, often in the form of starch or lipids. Plant components are consumed by animals and fungi, and used as fuel for cellular respiration. Oxidation of one gram of carbohydrate yields approximately 4 kcal of energy, while the oxidation of one gram of lipids yields about 9 kcal. Energy obtained from metabolism (e.g., oxidation of glucose) is usually stored temporarily within cells in the form of ATP. Organisms capable of aerobic respiration metabolize glucose and oxygen to release energy with carbon dioxide and water as byproducts.

Catabolism

Catabolism is the metabolic reaction which cells undergo to extract energy. There are two major metabolic pathways of monosaccharide catabolism: glycolysis and the citric acid cycle.

In glycolysis, oligo/polysaccharides are cleaved first to smaller monosaccharides by enzymes called glycoside hydrolases. The monosaccharide units can then enter into monosaccharide catabolism. In some cases, as with humans, not all carbohydrate types are usable as the digestive and metabolic enzymes necessary are not present.

Carbohydrate Chemistry

Carbohydrate chemistry is a large and economically important branch of organic chemistry. Some of the main organic reactions that involve carbohydrates are:

- Carbohydrate acetalisation
- Cyanohydrin reaction
- Lobry-de Bruyn-van Ekenstein transformation
- Amadori rearrangement
- Nef reaction
- Wohl degradation
- Koenigs–Knorr reaction
- Carbohydrate digestion

Lipid

Lipids are a group of naturally occurring molecules that include fats, waxes, sterols, fat-soluble vitamins (such as vitamins A, D, E, and K), monoglycerides, diglycerides, triglycerides, phospholipids, and others. The main biological functions of lipids include storing energy, signaling, and acting as structural components of cell membranes. Lipids have applications in the cosmetic and food industries as well as in nanotechnology.

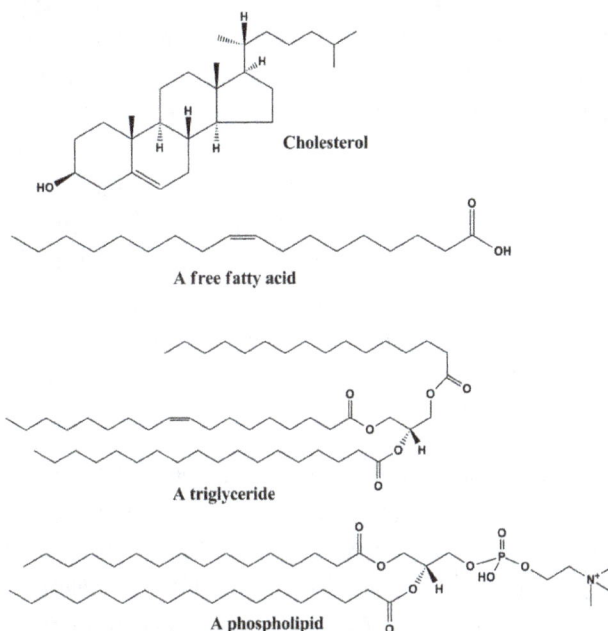

Structures of some common lipids. At the top are cholesterol and oleic acid. The middle structure is a triglyceride composed of oleoyl, stearoyl, andpalmitoyl chains attached to a glycerol backbone. At the bottom is the common phospholipid, phosphatidylcholine.

Lipids may be broadly defined as hydrophobic or amphiphilic small molecules; the amphiphilic nature of some lipids allows them to form structures such as vesicles, multilamellar/unilamellar liposomes, or membranes in an aqueous environment. Biological lipids originate entirely or in part from two distinct types of biochemical subunits or "building-blocks": ketoacyl and isoprene groups. Using this approach, lipids may be divided into eight categories: fatty acids, glycerolipids, glycerophospholipids, sphingolipids, saccharolipids, and polyketides (derived from condensation of ketoacyl subunits); and sterol lipids and prenol lipids (derived from condensation of isoprene subunits).

Although the term lipid is sometimes used as a synonym for fats, fats are a subgroup of lipids called triglycerides. Lipids also encompass molecules such as fatty acids and their derivatives (including tri-, di-, monoglycerides, and phospholipids), as well as other sterol-containing metabolites such as cholesterol. Although humans and other mammals use various biosynthetic pathways both to break down and to synthesize lipids, some essential lipids cannot be made this way and must be obtained from the diet.

Categories of Lipids

Fatty Acids

Fatty acids, or fatty acid residues when they are part of a lipid, are a diverse group of molecules synthesized by chain-elongation of an acetyl-CoA primer with malonyl-CoA or methylmalonyl-CoA groups in a process called fatty acid synthesis. They are made of a hydrocarbon chain that terminates with a carboxylic acid group; this arrangement confers the molecule with a polar, hydrophilic end, and a nonpolar, hydrophobic end that is insoluble in water. The fatty acid structure is one of the most fundamental categories of biological lipids, and is commonly used as a building-block of more structurally complex lipids. The carbon chain, typically between four and 24 carbons long, may be saturated or unsaturated, and may be attached to functional groups containing oxygen, halogens, nitrogen, and sulfur. If a fatty acid contains a double bond, there is the possibility of either a cis or trans geometric isomerism, which significantly affects the molecule's configuration. Cis-double bonds cause the fatty acid chain to bend, an effect that is compounded with more double bonds in the chain. Three double bonds in 18-carbon linolenic acid, the most abundant fatty-acyl chains of plant thylakoid membranes, render these membranes highly fluid despite environmental low-temperatures, and also makes linolenic acid give dominating sharp peaks in high resolution 13-C NMR spectra of chloroplasts. This in turn plays an important role in the structure and function of cell membranes. Most naturally occurring fatty acids are of the cis configuration, although the trans form does exist in some natural and partially hydrogenated fats and oils.

Examples of biologically important fatty acids include the eicosanoids, derived primarily from arachidonic acid and eicosapentaenoic acid, that include prostaglandins, leukotrienes, and thromboxanes. Docosahexaenoic acid is also important in biological systems, particularly with respect to sight. Other major lipid classes in the fatty acid category are the fatty esters and fatty amides. Fatty esters include important biochemical intermediates such as wax esters, fatty acid thioester coenzyme A derivatives, fatty acid thioester ACP derivatives and fatty acid carnitines. The fatty amides include N-acyl ethanolamines, such as the cannabinoid neurotransmitter anandamide.

Glycerolipids

Glycerolipids are composed of mono-, di-, and tri-substituted glycerols, the best-known being the fatty acid triesters of glycerol, called triglycerides. The word "triacylglycerol" is sometimes used synonymously with "triglyceride". In these compounds, the three hydroxyl groups of glycerol are each esterified, typically by different fatty acids. Because they function as an energy store, these lipids comprise the bulk of storage fat in animal tissues. The hydrolysis of the ester bonds of triglycerides and the release of glycerol and fatty acids from adipose tissue are the initial steps in metabolizing fat.

Additional subclasses of glycerolipids are represented by glycosylglycerols, which are characterized by the presence of one or more sugar residues attached to glycerol via a glycosidic linkage. Examples of structures in this category are the digalactosyldiacylglycerols found in plant membranes and seminolipid from mammalian sperm cells.

Glycerophospholipids

Glycerophospholipids, usually referred to as phospholipids, are ubiquitous in nature and are key components of the lipid bilayer of cells, as well as being involved in metabolism and cell signaling. Neural tissue (including the brain) contains relatively high amounts of glycerophospholipids, and alterations in their composition has been implicated in various neurological disorders. Glycerophospholipids may be subdivided into distinct classes, based on the nature of the polar headgroup at the sn-3 position of the glycerol backbone in eukaryotes and eubacteria, or the sn-1 position in the case of archaebacteria.

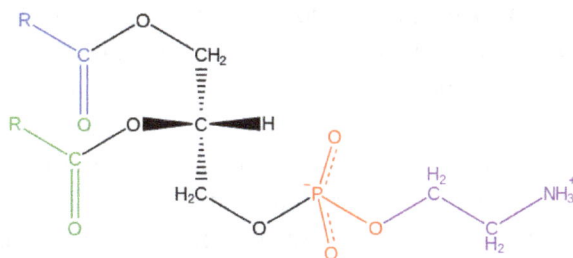

Phosphatidylethanolamine

Examples of glycerophospholipids found in biological membranes are phosphatidylcholine (also known as PC, GPCho or lecithin), phosphatidylethanolamine (PE or GPEtn) and phosphatidylserine (PS or GPSer). In addition to serving as a primary component of cellular membranes and binding sites for intra- and intercellular proteins, some glycerophospholipids in eukaryotic cells, such as phosphatidylinositols and phosphatidic acids are either precursors of or, themselves, membrane-derived second messengers. Typically, one or both of these hydroxyl groups are acylated with long-chain fatty acids, but there are also alkyl-linked and 1Z-alkenyl-linked (plasmalogen) glycerophospholipids, as well as dialkylether variants in archaebacteria.

Sphingolipids

Sphingolipids are a complicated family of compounds that share a common structural feature, a sphingoid base backbone that is synthesized de novo from the amino acid serine and a long-chain fatty acyl CoA, then converted into ceramides, phosphosphingolipids, glycosphingolipids and other compounds. The major sphingoid base of mammals is commonly referred to as sphingosine. Ceramides (N-acyl-sphingoid bases) are a major subclass of sphingoid base derivatives with an amide-linked fatty acid. The fatty acids are typically saturated or mono-unsaturated with chain lengths from 16 to 26 carbon atoms.

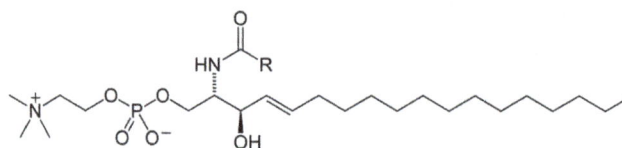

Sphingomyelin

The major phosphosphingolipids of mammals are sphingomyelins (ceramide phosphocholines), whereas insects contain mainly ceramide phosphoethanolamines and fungi have phytoceramide phosphoinositols and mannose-containing headgroups. The glycosphingolipids are a diverse family of molecules composed of one or more sugar residues linked via a glycosidic bond to the sphingoid base. Examples of these are the simple and complex glycosphingolipids such as cerebrosides and gangliosides.

Sterol lipids

Sterol lipids, such as cholesterol and its derivatives, are an important component of membrane lipids, along with the glycerophospholipids and sphingomyelins. The steroids, all derived from the same fused four-ring core structure, have different biological roles as hormones and signaling molecules. The eighteen-carbon (C18) steroids include the estrogen family whereas the C19 steroids comprise the androgens such as testosterone and androsterone. The C21 subclass includes the progestogens as well as the glucocorticoids and mineralocorticoids. The secosteroids, comprising various forms of vitamin D, are characterized by cleavage of the B ring of the core structure. Other examples of sterols are the bile acids and their conjugates, which in mammals are oxidized derivatives of cholesterol and are synthesized in the liver. The plant equivalents are the phytosterols, such as β-sitosterol, stigmasterol, and brassicasterol; the latter compound is also used as a biomarker for algal growth. The predominant sterol in fungal cell membranes is ergosterol.

Prenol Lipids

Prenol lipids are synthesized from the five-carbon-unit precursors isopentenyl diphosphate and dimethylallyl diphosphate that are produced mainly via the mevalonic acid (MVA) pathway. The simple isoprenoids (linear alcohols, diphosphates, etc.) are formed by the successive addition of C5 units, and are classified according to number of these terpene units. Structures containing greater than 40 carbons are known as polyterpenes. Carotenoids are important simple isoprenoids that function as antioxidants and as precursors of vitamin A. Another biologically important class of molecules is exemplified by the quinones and hydroquinones, which contain an isoprenoid tail attached to a quinonoid core of non-isoprenoid origin. Vitamin E and vitamin K, as well as the ubiquinones, are examples of this class. Prokaryotes synthesize polyprenols (called bactoprenols) in which the terminal isoprenoid unit attached to oxygen remains unsaturated, whereas in animal polyprenols (dolichols) the terminal isoprenoid is reduced.

Saccharolipids

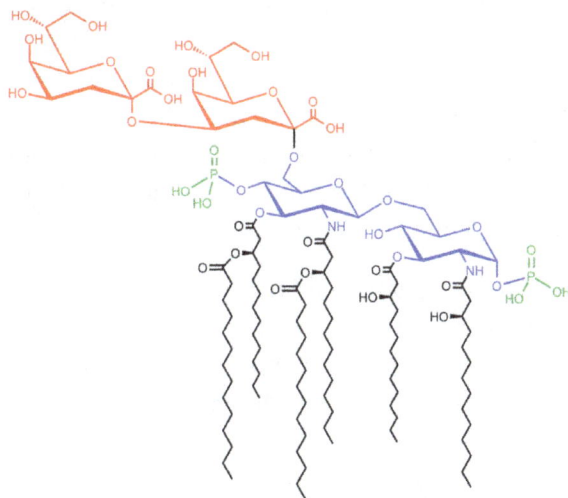

Structure of the saccharolipid Kdo$_2$-lipid A. Glucosamine residues in blue, Kdo residues in red, acyl chains in black and phosphate groups in green.

Saccharolipids describe compounds in which fatty acids are linked directly to a sugar backbone, forming structures that are compatible with membrane bilayers. In the saccharolipids, a monosaccharide substitutes for the glycerol backbone present in glycerolipids and glycerophospholipids. The most familiar saccharolipids are the acylated glucosamine precursors of the Lipid A component of the lipopolysaccharides in Gram-negative bacteria. Typical lipid A molecules are disaccharides of glucosamine, which are derivatized with as many as seven fatty-acyl chains. The minimal lipopolysaccharide required for growth in E. coli is Kdo$_2$-Lipid A, a hexa-acylated disaccharide of glucosamine that is glycosylated with two 3-deoxy-D-manno-octulosonic acid (Kdo) residues.

Polyketides

Polyketides are synthesized by polymerization of acetyl and propionyl subunits by classic enzymes as well as iterative and multimodular enzymes that share mechanistic features with the fatty acid synthases. They comprise a large number of secondary metabolites and natural products from animal, plant, bacterial, fungal and marine sources, and have great structural diversity. Many polyketides are cyclic molecules whose backbones are often further modified by glycosylation, methylation, hydroxylation, oxidation, and/or other processes. Many commonly used anti-microbial, anti-parasitic, and anti-cancer agents are polyketides or polyketide derivatives, such as erythromycins, tetracyclines, avermectins, and antitumor epothilones.

Biological Functions

Membranes

Eukaryotic cells feature compartmentalized membrane-bound organelles that carry out different biological functions. The glycerophospholipids are the main structural component of biological membranes, such as the cellular plasma membrane and the intracellular membranes of organelles; in animal cells the plasma membrane physically separates the intracellular components from the extracellular environment. The glycerophospholipids are amphipathic molecules (containing both hydrophobic and hydrophilic regions) that contain a glycerol core linked to two fatty acid-derived "tails" by ester linkages and to one "head" group by a phosphate ester linkage. While glycerophospholipids are the major component of biological membranes, other non-glyceride lipid components such as sphingomyelin and sterols (mainly cholesterol in animal cell membranes) are also found in biological membranes. In plants and algae, the galactosyldiacylglycerols, and sulfoquinovosyldiacylglycerol, which lack a phosphate group, are important components of membranes of chloroplasts and related organelles and are the most abundant lipids in photosynthetic tissues, including those of higher plants, algae and certain bacteria.

Plant thylakoid membranes have the largest lipid component of a non-bilayer forming monogalactosyl diglyceride (MGDG), and little phospholipids; despite this unique lipid composition, chloroplast thylakoid membranes have been shown to contain a dynamic lipid-bilayer matrix as revealed by magnetic resonance and electron microscope studies.

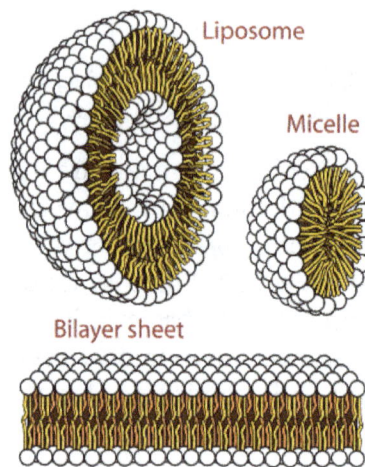

Self-organization of phospholipids: a spherical liposome, a micelle, and a lipid bilayer.

A biological membrane is a form of lamellar phase lipid bilayer. The formation of lipid bilayers is an energetically preferred process when the glycerophospholipids described above are in an aqueous environment. This is known as the hydrophobic effect. In an aqueous system, the polar heads of lipids align towards the polar, aqueous environment, while the hydrophobic tails minimize their contact with water and tend to cluster together, forming a vesicle; depending on the concentration of the lipid, this biophysical interaction may result in the formation of micelles, liposomes, or lipid bilayers. Other aggregations are also observed and form part of the polymorphism of amphiphile (lipid) behavior. Phase behavior is an area of study within biophysics and is the subject of current academic research. Micelles and bilayers form in the polar medium by a process known as the hydrophobic effect. When dissolving a lipophilic or amphiphilic substance in a polar environment, the polar molecules (i.e., water in an aqueous solution) become more ordered around the dissolved lipophilic substance, since the polar molecules cannot form hydrogen bonds to the lipophilic areas of the amphiphile. So in an aqueous environment, the water molecules form an ordered "clathrate" cage around the dissolved lipophilic molecule.

The formation of lipids into protocell membranes represents a key step in models of abiogenesis, the origin of life.

Energy Storage

Triglycerides, stored in adipose tissue, are a major form of energy storage both in animals and plants. The adipocyte, or fat cell, is designed for continuous synthesis and breakdown of triglycerides in animals, with breakdown controlled mainly by the activation of hormone-sensitive enzyme lipase. The complete oxidation of fatty acids provides high caloric content, about 9 kcal/g, compared with 4 kcal/g for the breakdown of carbohydrates and proteins. Migratory birds that must fly long distances without eating use stored energy of triglycerides to fuel their flights.

Signaling

In recent years, evidence has emerged showing that lipid signaling is a vital part of the cell signaling. Lipid signaling may occur via activation of G protein-coupled or nuclear receptors, and members of several different lipid categories have been identified as signaling molecules and cellular messengers. These include sphingosine-1-phosphate, a sphingolipid derived from ceramide that is a potent messenger molecule involved in regulating calcium mobilization, cell growth, and apoptosis; diacylglycerol (DAG) and the phosphatidylinositol phosphates (PIPs), involved in calcium-mediated activation of protein kinase C; the prostaglandins, which are one type of fatty-acid derived eicosanoid involved in inflammation and immunity; the steroid hormones such as estrogen, testosterone and cortisol, which modulate a host of functions such as reproduction, metabolism and blood pressure; and the oxysterols such as 25-hydroxy-cholesterol that are liver X receptor agonists. Phosphatidylserine lipids are known to be involved in signaling for the phagocytosis of apoptotic cells and/or pieces of cells. They accomplish this by being exposed to the extracellular face of the cell membrane after the inactivation of flippases which place them exclusively on the cytosolic side and the activation of scramblases, which scramble the orientation of the phospholipids. After this occurs, other cells recognize the phosphatidylserines and phagocytosize the cells or cell fragments exposing them.

Other Functions

The "fat-soluble" vitamins (A, D, E and K) – which are isoprene-based lipids – are essential nutrients stored in the liver and fatty tissues, with a diverse range of functions. Acyl-carnitines are involved in the transport and metabolism of fatty acids in and out of mitochondria, where they undergo beta oxidation. Polyprenols and their phosphorylated derivatives also play important transport roles, in this case the transport of oligosaccharides across membranes. Polyprenol phosphate sugars and polyprenol diphosphate sugars function in extra-cytoplasmic glycosylation reactions, in extracellular polysaccharide biosynthesis (for instance, peptidoglycan polymerization in bacteria), and in eukaryotic protein N-glycosylation. Cardiolipins are a subclass of glycerophospholipids containing four acyl chains and three glycerol groups that are particularly abundant in the inner mitochondrial membrane. They are believed to activate enzymes involved with oxidative phosphorylation. Lipids also form the basis of steroid hormones.

Metabolism

The major dietary lipids for humans and other animals are animal and plant triglycerides, sterols, and membrane phospholipids. The process of lipid metabolism synthesizes and degrades the lipid stores and produces the structural and functional lipids characteristic of individual tissues.

Biosynthesis

In animals, when there is an oversupply of dietary carbohydrate, the excess carbohydrate is converted to triglycerides. This involves the synthesis of fatty acids from acetyl-CoA and the esterification of fatty acids in the production of triglycerides, a process called lipogenesis. Fatty acids are made by fatty acid synthases that polymerize and then reduce acetyl-CoA units. The acyl chains in the fatty acids are extended by a cycle of reactions that add the acetyl group, reduce it to an alcohol, dehydrate it to an alkene group and then reduce it again to an alkane group. The enzymes of fatty acid biosynthesis are divided into two groups, in animals and fungi all these fatty acid synthase reactions are carried out by a single multifunctional protein, while in plant plastids and bacteria separate enzymes perform each step in the pathway. The fatty acids may be subsequently converted to triglycerides that are packaged in lipoproteins and secreted from the liver.

The synthesis of unsaturated fatty acids involves a desaturation reaction, whereby a double bond is introduced into the fatty acyl chain. For example, in humans, the desaturation of stearic acid by stearoyl-CoA desaturase-1 produces oleic acid. The doubly unsaturated fatty acid linoleic acid as well as the triply unsaturated α-linolenic acid cannot be synthesized in mammalian tissues, and are therefore essential fatty acids and must be obtained from the diet.

Triglyceride synthesis takes place in the endoplasmic reticulum by metabolic pathways in which acyl groups in fatty acyl-CoAs are transferred to the hydroxyl groups of glycerol-3-phosphate and diacylglycerol.

Terpenes and isoprenoids, including the carotenoids, are made by the assembly and modification of isoprene units donated from the reactive precursors isopentenyl pyrophosphate and dimethylallyl pyrophosphate. These precursors can be made in different ways. In animals and archaea, the mevalonate pathway produces these compounds from acetyl-CoA, while in plants and bacteria the non-mevalonate pathway uses pyruvate and glyceraldehyde 3-phosphate as substrates. One important reaction that uses these activated isoprene donors is steroid biosynthesis. Here, the isoprene units are joined together to make squalene and then folded up and formed into a set of rings to make lanosterol. Lanosterol can then be converted into other steroids such as cholesterol and ergosterol.

Degradation

Beta oxidation is the metabolic process by which fatty acids are broken down in the mitochondria and/or in peroxisomes to generate acetyl-CoA. For the most part, fatty acids are oxidized by a mechanism that is similar to, but not identical with, a reversal of the process of fatty acid synthesis. That is, two-carbon fragments are removed sequentially from the carboxyl end of the acid after steps of dehydrogenation, hydration, and oxidation to form a beta-keto acid, which is split by thiolysis. The acetyl-CoA is then ultimately converted into ATP, CO_2, and H_2O using the citric acid cycle and the electron transport chain. Hence the citric acid cycle can start at acetyl-CoA when fat is being broken down for energy if there is little or no glucose available. The energy yield of the complete oxidation of the fatty acid palmitate is 106 ATP. Unsaturated and odd-chain fatty acids require additional enzymatic steps for degradation.

Nutrition and Health

Most of the fat found in food is in the form of triglycerides, cholesterol, and phospholipids. Some dietary fat is necessary to facilitate absorption of fat-soluble vitamins (A, D, E, and K) and carotenoids. Humans and other mammals have a dietary requirement for certain essential fatty acids, such as linoleic acid (an omega-6 fatty acid) and alpha-linolenic acid (an omega-3 fatty acid) because they cannot be synthesized from simple precursors in the diet. Both of these fatty acids are 18-carbon polyunsaturated fatty acids differing in the number and position of the double bonds. Most vegetable oils are rich in linoleic acid (safflower, sunflower, and corn oils). Alpha-linolenic acid is found in the green leaves of plants, and in selected seeds, nuts, and legumes (in particular flax, rapeseed, walnut, and soy). Fish oils are particularly rich in the longer-chain omega-3 fatty acids eicosapentaenoic acid (EPA) and docosahexaenoic acid (DHA). A large number of studies have shown positive health benefits associated with consumption of omega-3 fatty acids on infant development, cancer, cardiovascular diseases, and various mental illnesses, such as depression, attention-deficit hyperactivity disorder, and dementia. In contrast, it is now well-established that consumption of trans fats, such as those present in partially hydrogenated vegetable oils, are a risk factor for cardiovascular disease.

A few studies have suggested that total dietary fat intake is linked to an increased risk of obesity and diabetes. However, a number of very large studies, including the Women's Health Initiative Dietary Modification Trial, an eight-year study of 49,000 women, the Nurses' Health Study and the Health Professionals Follow-up Study, revealed no such links. None of these studies suggested any connection between percentage of calories from fat and risk of cancer, heart disease, or weight gain. The Nutrition Source, a website maintained by the Department of Nutrition at the Harvard School of Public Health, summarizes the current evidence on the impact of dietary fat: "Detailed research—much of it done at Harvard—shows that the total amount of fat in the diet isn't really linked with weight or disease."

Further Classification of Lipids

Lipid Microdomain

Often, lateral heterogeneity has been inferred from biophysical techniques where the observed signal indicates multiple populations rather than the expected homogenous population. An example of this is the measurement of the diffusion coefficient of a fluorescent lipid analogue in soybean protoplasts. Membrane microheterogeneity is sometimes inferred from the behavior of enzymes, where the enzymatic activity does not appear to be correlated with the average lipid physical state exhibited by the bulk of the membrane. Often, the methods suggest regions with different lipid fluidity, as would be expected of coexisting gel and liquid crystalline phases within the biomembrane. This is also the conclusion of a series of studies where differential effects of perturbation caused by cis and trans fatty acids are interpreted in terms of preferential partitioning of the two liquid crystalline and gel-like domains.

Lipid Signaling

Lipid signaling, broadly defined, refers to any biological signaling event involving a lipid messenger that binds a protein target, such as a receptor, kinase or phosphatase, which in turn mediate the effects of these lipids on specific cellular responses. Lipid signaling is thought to be qualitatively different from other classical signaling paradigms (such as monoamine neurotransmission) because lipids can freely diffuse through membranes. One consequence of this is that lipid messengers cannot be stored in vesicles prior to release and so are often biosynthesized "on demand" at their intended site of action. As such, many lipid signaling molecules cannot circulate freely in solution but, rather, exist bound to special carrier proteins in serum.

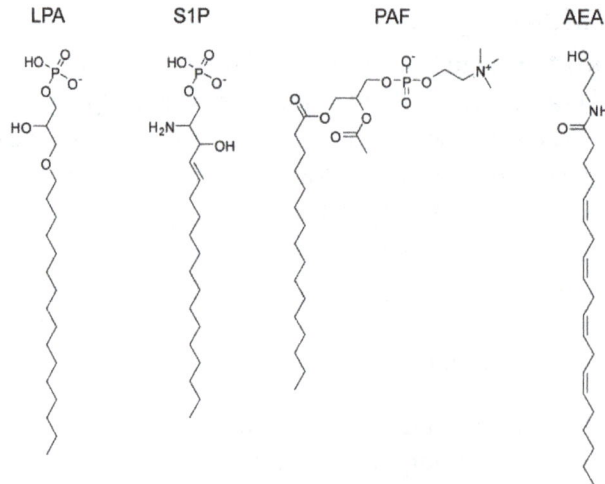

Common lipid signaling molecules:
lysophosphatidic acid (LPA)
sphingosine-1-phosphate (S1P)
platelet activating factor (PAF)
anandamide or arachidonoyl ethanolamine (AEA)

Sphingolipid Second Messengers

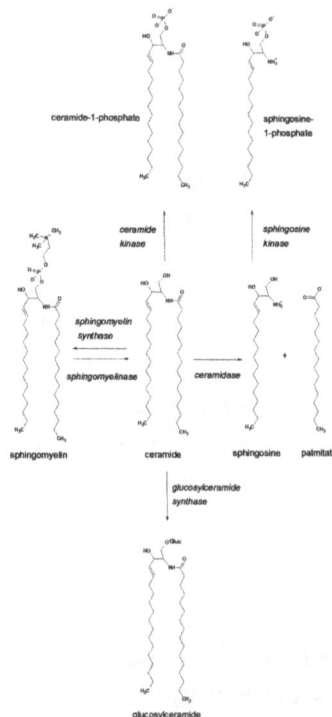

Sphingolipid second messengers. Ceramide is at the metabolic hub, leading to the formation of other sphingolipids.

Ceramide

Ceramide (Cer) can be generated by the breakdown of sphingomyelin (SM) by sphingomyelinases (SMases), which are enzymes that hydrolyze the phosphocholine group from the sphingosine backbone. Alternatively, this sphingosine-derived lipid (sphingolipid) can be synthesized from scratch (de novo) by the enzymes serine palmitoyl transferase (SPT) and ceramide synthase in organelles such as the endoplasmic reticulum (ER) and possibly, in the mitochondria-associated membranes (MAMs) and the perinuclear membranes. Being located in the metabolic hub, ceramide leads to the formation of other sphingolipids, with the C_1 hydroxyl (-OH) group as the major site of modification. A sugar can be attached to ceramide (glycosylation) through the action of the enzymes, glucosyl or galactosyl ceramide synthases. Ceramide can also be broken down by enzymes called ceramidases, leading to the formation of sphingosine, Moreover, a phosphate group can be attached to ceramide (phosphorylation) by the enzyme, ceramide kinase. It is also possible to regenerate sphingomyelin from ceramide by accepting a phosphocholine headgroup from phosphatidylcholine (PC) by the action of an enzyme called sphingomyelin synthase. The latter process results in the formation of diacylglycerol (DAG) from PC.

Ceramide contains two hydrophobic ("water-fearing") chains and a neutral headgroup. Consequently, it has limited solubility in water and is restricted within the organelle where it was formed. Also, because of its hydrophobic nature, ceramide readily flip-flops across membranes as supported by studies in membrane models and membranes from red blood cells (erythrocytes). However, ceramide can possibly interact with other lipids to form bigger regions called microdomains which restrict its flip-flopping abilities. This could have immense effects on the signaling functions of ceramide because it is known that ceramide generated by acidic SMase enzymes in the outer leaflet of an organelle membrane may have different roles compared to ceramide that is formed in the inner leaflet by the action of neutral SMase enzymes.

Ceramide mediates many cell-stress responses, including the regulation of programmed cell death (apoptosis) and cell aging (senescence). Numerous research works have focused interest on defining the direct protein targets of action of ceramide. These include enzymes called ceramide-activated Ser-Thr phosphatases (CAPPs), such as protein phosphatase 1 and 2A (PP1 and PP2A), which were found to interact with ceramide in studies done in a controlled environment outside of a living organism (in vitro). On the other hand, studies in cells have shown that ceramide-inducing agents such as tumor necrosis factor-alpha α (TNFα) and palmitate induce the ceramide-dependent removal of a phosphate group (dephosphorylation) of the retinoblastoma gene product RB and the enzymes, protein kinases B (AKT protein family) and C α (PKB and PKCα). Moreover, there is also sufficient evidence which implicates ceramide to the activation of the kinase suppressor of Ras (KSR), PKCζ, and cathepsin D. Interestingly, cathepsin D has been proposed as the main target for ceramide formed in organelles called lysosomes, making lysosomal acidic SMase enzymes one of the key players in the mitochondrial pathway of apoptosis. Ceramide was also shown to activate PKCζ, implicating it to the inhibition of AKT, regulation of the voltage difference between the interior and exterior of the cell (membrane potential) and signaling functions that favor apoptosis. Chemotherapeutic agents such as daunorubicin and etoposide enhance the de novo synthesis of ceramide in studies done on mammalian cells. The same results were found for certain inducers of apoptosis particularly stimulators of receptors in a class of lymphocytes (a type of white blood cell) called B-cells. Regulation of the de novo synthesis of ceramide by palmitate may have a key role in diabetes and the metabolic syndrome. Experimental evidence shows that there is substantial increase of ceramide levels upon adding palmitate. Ceramide accumulation activates PP2A and the subsequent dephosphorylation and inactivation of AKT, a crucial mediator in metabolic control and insulin signaling. This results in a substantial decrease in insulin responsiveness (i.e. to glucose) and in the death of insulin-producing cells in the pancreas called islets of Langerhans. Inhibition of ceramide synthesis in mice via drug treatments or gene-knockout techniques prevented insulin resistance induced by fatty acids, glucocorticoids or obesity.

An increase in in vitro activity of acid SMase has been observed after applying multiple stress stimuli such as ultraviolet (UV) and ionizing radiation, binding of death receptors and chemotherapeutic agents such as platinum, histone deacetylase inhibitors and paclitaxel. In some studies, SMase activation results to its transport to the plasma membrane and the simultaneous formation of ceramide.

Ceramide transfer protein (CERT) transports ceramide from ER to the Golgi for the synthesis of SM. CERT is known to bind phosphatidylinositol phosphates, hinting its potential regulation via phosphorylation, a step of the ceramide metabolism that can be enzymatically regulated by protein kinases and phosphatases, and by inositol lipid metabolic pathways. Up to date, there are at least 26 distinct enzymes with varied subcellular localizations, that act on ceramide as either a substrate or product. Regulation of ceramide levels can therefore be performed by one of these enzymes in distinct organelles by particular mechanisms at various times.

Sphingosine

Sphingosine (Sph) is formed by the action of ceramidase (CDase) enzymes on ceramide in the lysosome. Sph can also be formed in the extracellular (outer leaflet) side of the plasma membrane by the action of neutral CDase enzyme. Sph then is either recycled back to ceramide or phosphorylated by one of the sphingosine kinase enzymes, SK1 and SK2. The product sphingosine-1-phosphate (S1P) can be dephosphorylated in the ER to regenerate sphingosine by certain S1P phosphatase enzymes within cells, where the salvaged Sph is recycled to ceramide. Sphingosine is a single-chain lipid (usually 18 carbons in length), rendering it to have sufficient solubility in water. This explains its ability to move between membranes and to flip-flop across a membrane. Estimates conducted at physiological pH show that approximately 70% of sphingosine remains in membranes while the remaining 30% is water-soluble. Sph that is formed has sufficient solubility in the liquid found inside cells (cytosol). Thus, Sph may come out of the lysosome and move to the ER without the need for transport via proteins or membrane-enclosed sacs called vesicles. However, its positive charge favors partitioning in lysosomes. It is proposed that the role of SK1 located near or in the lysosome is to 'trap' Sph via phosphorylation.

It is important to note that since sphingosine exerts surfactant activity, it is one of the sphingolipids found at lowest cellular levels. The low levels of Sph and their increase in response to stimulation of cells, primarily by activation of ceramidase by growth-inducing proteins such as platelet-derived growth factor and insulin-like growth factor, is consistent with its function as a second messenger. It was found that immediate hydrolysis of only 3 to 10% of newly generated ceramide may double the levels of Sph. Treatment of HL60 cells (a type of leukemia cell line) by a plant-derived organic compound called phorbol ester increased Sph levels threefold, whereby the cells differentiated into white blood cells called macrophages. Treatment of the same cells by exogenous Sph caused apoptosis. A specific protein kinase phosphorylates 14-3-3, otherwise known as sphingosine-dependent protein kinase 1 (SDK1), only in the presence of Sph.

Sph is also known to interact with protein targets such as the protein kinase H homologue (PKH) and the yeast protein kinase (YPK). These targets in turn mediate the effects of Sph and its related sphingoid bases, with known roles in regulating the actin cytoskeleton, endocytosis, the cell cycle and apoptosis. It is important to note however that the second messenger function of Sph is not yet established unambiguously.

Sphingosine-1-Phosphate

Sphingosine-1-phosphate (S1P), like Sph, is composed of a single hydrophobic chain and has sufficient solubility to move between membranes. S1P is formed by phosphorylation of sphingosine by sphingosine kinase (SK). The phosphate group of the product can be detached (dephosphorylated) to regenerate sphingosine via S1P phosphatase enzymes or S1P can be broken down by S1P lyase enzymes to ethanolamine phosphate and hexadecenal. Similar to Sph, its second messenger function is not yet clear. However, there is substantial evidence that implicates S1P to cell survival, cell migration, and inflammation. Certain growth-inducing proteins such as platelet-derived growth factor (PDGF), insulin-like growth factor (IGF) and vascular endothelial growth factor (VEGF) promote the formation of SK enzymes, leading to increased levels of S1P. Other factors that induce SK include cellular communication molecules called cytokines, such as tumor necrosis factor α (TNFα) and interleukin-1 (IL-1), hypoxia or lack of oxygen supply in cells, oxidized low-density lipoproteins (oxLDL) and several immune complexes.

S1P is probably formed at the inner leaflet of the plasma membrane in response to TNFα and other receptor activity-altering compounds called agonists. S1P, being present in low nanomolar concentrations in the cell, has to interact with high-affinity receptors that are capable of sensing their low levels. So far, the only identified receptors for S1P are the high-affinity G protein-coupled receptors (GPCRs), also known as S1P receptors (S1PRs). S1P is required to reach the extracellular side (outer leaflet) of the plasma membrane to interact with S1PRs and launch typical GPCR signaling pathways. However, the zwitterionic headgroup of S1P makes it unlikely to flip-flop spontaneously. To overcome this difficulty, the ATP-binding cassette (ABC) transporter C1 (ABCC1) serves as the "exit door" for S1P. On the other hand, the cystic fibrosis transmembrane regulator (CFTR) serves as the means of entry for S1P into the cell. In contrast to its low intracellular concentration, S1P is found in high nanomolar concentrations in serum where it is bound to albumin and lipoproteins. Inside the cell, S1P can induce calcium release independent of the S1PRs—the mechanism of which remains unknown. To date, the intracellular molecular targets for S1P are still unidentified.

The SK1-S1P pathway has been extensively studied in relation to cytokine action, with multiple functions connected to effects of TNFα and IL-1 favoring inflammation. Studies show that knock-down of key enzymes such as S1P lyase and S1P phosphatase increased prostaglandin production, parallel to increase of S1P levels. This strongly suggests that S1P is the mediator of SK1 action and not subsequent compounds. Research done on endothelial and smooth muscle cells is consistent to the hypothesis that S1P has a crucial role in regulating endothelial cell growth, and movement. Recent work on a sphingosine analogue, FTY270, demonstrates its ability to act as a potent compound that alters the activity of S1P receptors (agonist). FTY270 was further verified in clinical tests to have roles in immune modulation, such as that on multiple sclerosis. This highlights the importance of S1P in the regulation of lymphocyte function and immunity. Most of the studies on S1P are used to further understand diseases such as cancer, arthritis and inflammation, diabetes, immune function and neurodegenerative disorders.

Glucosylceramide

Glucosylceramides (GluCer) are the most widely distributed glycosphingolipids in cells serving as precursors for the formation of over 200 known glycosphingolipids. GluCer is formed by the glycosylation of ceramide in an organelle called Golgi via enzymes called glucosylceramide synthase (GCS) or by the breakdown of complex glycosphingolipids (GSLs) through the action of specific hydrolase enzymes. In turn, certain β-glucosidases hydrolyze these lipids to regenerate ceramide. GluCer appears to be synthesized in the inner leaflet of the Golgi. Studies show that GluCer has to flip to the inside of the Golgi or transfer to the site of GSL synthesis to initiate the synthesis of complex GSLs. Transferring to the GSL synthesis site is done with the help of a transport protein known as four phosphate adaptor protein 2 (FAPP2) while the flipping to the inside of the Golgi is made possible by the ABC transporter P-glycoprotein, also known as the multi-drug resistance 1 transporter (MDR1). GluCer is implicated in post-Golgi trafficking and drug resistance particularly to chemotherapeutic agents. For instance, a study demonstrated a correlation between cellular drug resistance and modifications in GluCer metabolism.

In addition to their role as building blocks of biological membranes, glycosphingolipids have long attracted attention because of their supposed involvement in cell growth, differentiation, and formation of tumors. The production of GluCer from Cer was found to be important in the growth of neurons or brain cells. On the other hand, pharmacological inhibition of GluCer synthase is being considered a technique to avoid insulin resistance.

Ceramide-1-Phosphate

Ceramide-1-phosphate (C1P) is formed by the action of ceramide kinase (CK) enzymes on Cer. C1P carry ionic charge at neutral pH and contain two hydrophobic chains making it relatively insoluble in aqueous environment. Thus, C1P reside in the organelle where it was formed and is unlikely to spontaneously flip-flop across membrane bilayers.

C1P activate phospholipase A2 and is found, along with CK, to be a mediator of arachidonic acid released in cells in response to a protein called interleukin-1β (IL-1β) and a lipid-soluble molecule that transports calcium ions (Ca^+) across the bilayer, also known as calcium ionophore. C1P was also previously reported to encourage cell division (mitogenic) in fibroblasts, block apoptosis by inhibiting acid SMase in white blood cells within tissues (macrophages) and increase intracellular free calcium concentrations in thyroid cells. C1P also has known roles in vesicular trafficking, cell survival, phagocytosis ("cell eating") and macrophage degranulation.

Phosphatidylinositol Bisphosphate (PIP_2) Lipid Agonist

PIP_2 binds directly to ion channels and modulates their activity. PIP_2 was shown to directly agonizes Inward rectifying potassium channels(K_{ir}). In this regard intact PIP_2 signals as a bona fide neurotransmitter-like ligand. PIP_2's interaction with many ion channels suggest that the intact form of PIP_2 has an important signaling role independent of second messenger signaling.

Second Messengers from Phosphatidylinositol

Phosphatidylinositol Bisphosphate (Pip$_2$) Second Messenger Systems

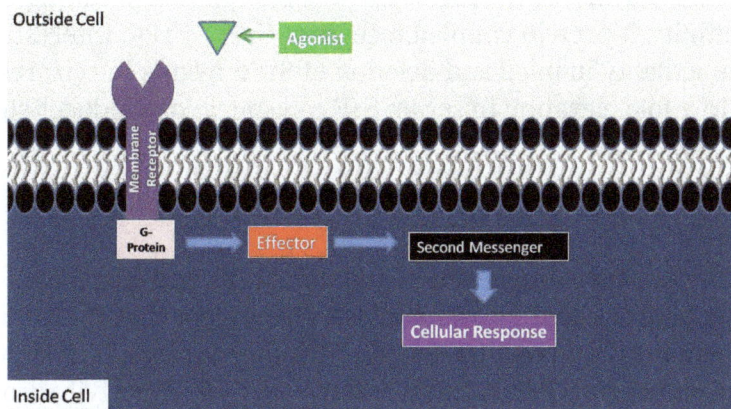

Cartoon of second messenger systems. Figure adapted From Barbraham Institute Mike Berridge.

A general second messenger system mechanism can be broken down into four steps. First, the agonist activates a membrane-bound receptor. Second, the activated G-protein produces a primary effector. Third, the primary effect stimulates the second messenger synthesis. Fourth, the second messenger activates a certain cellular process.

The G-protein coupled receptors for the PIP$_2$ messenger system produces two effectors, phospholipase C (PLC) and phosphoinositide 3-kinase (PI3K). PLC as an effector produces two different second messengers, inositol triphosphate (IP$_3$) and Diacylglycerol (DAG).

IP$_3$ is soluble and diffuses freely into the cytoplasm. As a second messenger, it is recognized by the inositol triphosphate receptor (IP3R), a Ca$^+$ channel in the endoplasmic reticulum (ER) membrane, which stores intracellular Ca$^+$. The binding of IP$_3$ to IP3R releases Ca$^+$ from the ER into the normally Ca$^+$-poor cytoplasm, which then triggers various events of Ca$^+$ signaling. Specifically in blood vessels, the increase in Ca$^+$ concentration from IP$_3$ releases nitric oxide, which then diffuses into the smooth muscle tissue and causes relaxation.

DAG remains bound to the membrane by its fatty acid "tails" where it recruits and activates both conventional and novel members of the protein kinase C family. Thus, both IP$_3$ and DAG contribute to activation of PKCs.

Phosphoinositide 3-kinase (PI3K) as an effector phosphorylates phosphatidylinositol bisphosphate (PIP$_2$) to produce phosphatidylinositol (3,4,5)-trisphosphate (PIP$_3$). PIP$_3$ has been shown to activate protein kinase B, increase binding to extracellular proteins and ultimately enhance cell survival.

Activators of G-Protein Coupled Receptors

Lysophosphatidic Acid (LPA)

LPA is the result of phospholipase A2 action on phosphatidic acid. The SN-1 position can contain either an ester bond or an ether bond, with ether LPA being found at elevated levels in certain cancers. LPA binds the high-affinity G-protein coupled receptors LPA1, LPA2, and LPA3 (also known as EDG2, EDG4, and EDG7, respectively).

Sphingosine-1-Phosphate (S1P)

S1P is present at high concentrations in plasma and secreted locally at elevated concentrations at sites of inflammation. It is formed by the regulated phosphorylation of sphingosine. It acts through five dedicated high-affinity G-protein coupled receptors, S1P1 - S1P5. Interestingly, targeted deletion of S1P1 results in lethality in mice and deletion of S1P2 results in seizures and deafness. Additionally, a mere 3- to 5-fold elevation in serum S1P concentrations induces sudden cardiac death by an S1P3-receptor specific mechanism.

Platelet Activating Factor (PAF)

PAF is a potent activator of platelet aggregation, inflammation, and anaphylaxis. It is similar to the ubiquitous membrane phospholipid phosphatidylcholine except that it contains an acetyl-group in the SN-2 position and the SN-1 position contains an ether-linkage. PAF signals through a dedicated G-protein coupled receptor, PAFR and is inactivated by PAF acetylhydrolase.

Endocannabinoids

The endogenous cannabinoids, or endocannabinoids, are endogenous lipids that activate cannabinoid receptors. The first such lipid to be isolated was anandamide which is the arachidonoyl amide of ethanolamine. Anandamide is formed via enzymatic release from N-arachidonoyl phosphatidylethanolamine by the N-acyl phosphatidylethanolamine phospholipase D (NAPE-PLD). Anandamide activates both the CB1 receptor, found primarily in the central nervous system, and the CB2 receptor which is found primarily in lymphocytes and the periphery. It is found at very low levels (nM) in most tissues and is inactivated by the fatty acid amide hydrolase. Subsequently, another endocannabinoid was isolated, 2-arachidonoylglycerol, which is produced when phospholipase C releases diacylglycerol which is then converted to 2-AG by diacylglycerol lipase. 2-AG can also activate both cannabinoid receptors and is inactivated by monoacylglycerol lipase. It is present at approximately 100-times the concentration of anandamide in most tissues. Elevations in either of these lipids causes analgesia and anti-inflammation and tissue protection during states of ischemia, but the precise roles played by these various endocannabinoids are still not totally known and intensive research into their function, metabolism, and regulation is ongoing. One saturated lipid from this class, often called an endocannabinoid, but with no relevant affinity for the CB1 and CB 2 receptor is palmitoylethanolamide. This signaling lipid has great affinity for the GRP55 receptor and the PPAR alpha receptor. It has been identified as an anti-inflammatory compound already in 1957, and as an analgesic compound in 1975. It was Rita Levi-Montalcini. who first identified one of its biological mechanisms of action, the inhibition of activated mast cells. Palmitoylethanolamide is the only endocannabinoid available on the market for treatment, as a food supplement.

Prostaglandins

Prostaglandins are formed through oxidation of arachidonic acid by cyclooxygenases and other prostaglandin synthases. There are currently nine known G-protein coupled receptors (eicosanoid receptors) that largely mediate prostaglandin physiology.

FAHFA

FAHFAs (fatty acid esters of hydroxy fatty acids) are formed in adipose tissue, improve glucose tolerance and also reduce adipose tissue inflammation. Palmitic acid esters of hydroxy-stearic acids (PAHSAs) are among the most bioactive members able to activate G-protein coupled receptors 120. Docosahexaenoic acid ester of hydroxy-linoleic acid (DHAHLA) exert anti-inflammatory and pro-resolving properties.

Retinol Derivatives

Retinaldehyde is a retinol (vitamin A) derivative responsible for vision. It binds rhodopsin, a well-characterized GPCR that binds all-cis retinal in its inactive state. Upon photoisomerization by a photon the cis-retinal is converted to trans-retinal causing activation of rhodopsin which ultimately leads to depolarization of the neuron thereby enabling visual perception.

Activators of Nuclear Receptors

Steroid Hormones

This large and diverse class of steroids are biosynthesized from isoprenoids and structurally resemble cholesterol. Mammalian steroid hormones can be grouped into five groups by the receptors to which they bind: glucocorticoids, mineralocorticoids, androgens, estrogens, and progestogens.

Retinoic Acid

Retinol (vitamin A) can be metabolized to retinoic acid which activates nuclear receptors such as the RAR to control differentiation and proliferation of many types of cells during development.

Prostaglandins

The majority of prostaglandin signaling occurs via GPCRs although certain prostaglandins activate nuclear receptors in the PPAR family.

Lipidomics

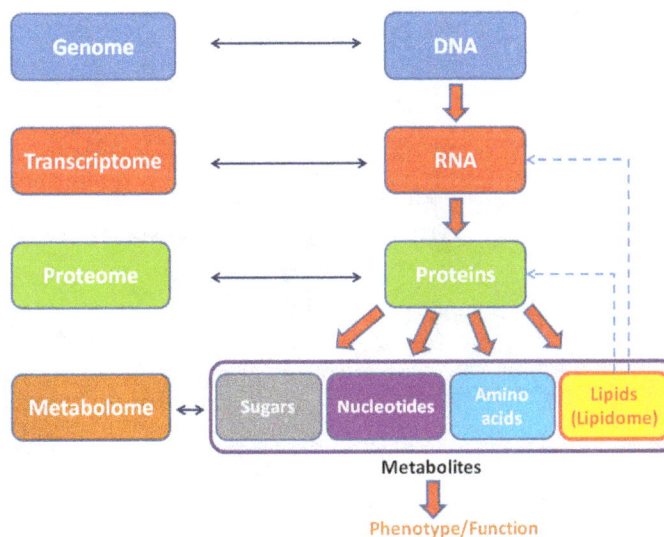

General schema showing the relationships of the lipidome to the genome, transcriptome, proteome and metabolome. Lipids also regulate protein function and gene transcription as part of a dynamic "interactome" within the cell.

Lipidomics is the large-scale study of pathways and networks of cellular lipids in biological systems The word "lipidome" is used to describe the complete lipid profile within a cell, tissue or organism and is a subset of the "metabolome" which also includes the three other major classes of biological molecules: proteins/amino-acids, sugars and nucleic acids. Lipidomics is a relatively recent research field that has been driven by rapid advances in technologies such as mass spectrometry (MS), nuclear magnetic resonance (NMR) spectroscopy, fluorescence spectroscopy, dual polarisation interferometry and computational methods, coupled with the recognition of the role of lipids in many metabolic diseases such as obesity, atherosclerosis, stroke, hypertension and diabetes. This rapidly expanding field complements the huge progress made in genomics and proteomics, all of which constitute the family of systems biology.

Lipidomics research involves the identification and quantification of the thousands of cellular lipid molecular species and their interactions with other lipids, proteins, and other metabolites. Investigators in lipidomics examine the structures, functions, interactions, and dynamics of cellular lipids and the changes that occur during perturbation of the system.

Han and Gross first defined the field of lipidomics through integrating the specific chemical properties inherent in lipid molecular species with a comprehensive mass spectrometric approach. Although lipidomics is under the umbrella of the more general field of "metabolomics", lipidomics is itself a distinct discipline due to the uniqueness and functional specificity of lipids relative to other metabolites.

In lipidomic research, a vast amount of information quantitatively describing the spatial and temporal alterations in the content and composition of different lipid molecular species is accrued after perturbation of a cell through changes in its physiological or pathological state. Information obtained from these studies facilitates mechanistic insights into changes in cellular function. Therefore, lipidomic studies play an essential role in defining the biochemical mechanisms of lipid-related disease processes through identifying alterations in cellular lipid metabolism, trafficking and homeostasis. The growing attention on lipid research is also seen from the initiatives underway of the LIPID Metabolites And Pathways Strategy (LIPID MAPS Consortium). and The European Lipidomics Initiative (ELIfe).

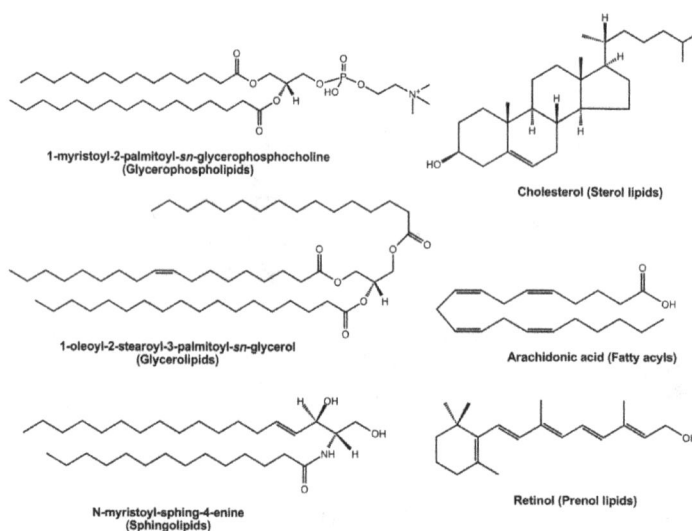

Examples of some lipids from various categories.

Structural Diversity of Lipids

Lipids are a diverse and ubiquitous group of compounds which have many key biological functions, such as acting as structural components of cell membranes, serving as energy storage sources and participating in signaling pathways. Lipids may be broadly defined as hydrophobic or amphipathic small molecules that originate entirely or in part from two distinct types of biochemical subunits or "building blocks": ketoacyl and isoprene groups. The huge structural diversity found in lipids arises from the biosynthesis of various combinations of these building blocks. For example, glycerophospholipids are composed of a glycerol backbone linked to one of approximately 10 possible headgroups and also to 2 fatty acyl/alkyl chains, which in turn may have 30 or more different molecular structures. In practice, not all possible permutations are detected experimentally, due to chain preferences depending on the cell type and also to detection limits - nevertheless several hundred distinct glycerophospholipid molecular species have been detected in mammalian cells.

Plant chloroplast thylakoid membranes however, have unique lipid composition as they are deficient in phospholipids. Also, their largest constituent, monogalactosyl diglyceride or MGDG, does not form aqueous bilayers. Nevertheless, dyamic studies reveal a normal lipid bilayer organisation in thylakoid membranes.

Experimental Techniques

Lipid Extraction

Most methods of lipid extraction and isolation from biological samples exploit the high solubility of hydrocarbon chains in organic solvents. Given the diversity in lipid classes, it is not possible to accommodate all classes with a common extraction method. The traditional Bligh/Dyer procedure uses chloroform/methanol-based protocols that include phase partitioning into the organic layer. These protocols work relatively well for a wide variety of physiologically relevant lipids but they have to be adapted for complex lipid chemistries and low-abundance and labile lipid metabolites. When organic soil was used, citrate buffer in the extraction mixture gave higher amounts of lipid phosphate than acetate buffer, Tris, H_2O or phosphate buffer.

Lipid Separation

The simplest method of lipid separation is the use of thin layer chromatography (TLC). Although not as sensitive as other methods of lipid detection, it offers a rapid and comprehensive screening tool prior to more sensitive and sophisticated techniques. Solid-phase extraction (SPE) chromatography is useful for rapid, preparative separation of crude lipid mixtures into different lipid classes. This involves the use of prepacked columns containing silica or other stationary phases to separate glycerophospholipids, fatty acids, cholesteryl esters, glycerolipids, and sterols from crude lipid mixtures. High-performance liquid chromatography (HPLC or LC) is extensively used in lipidomic analysis to separate lipids prior to mass analysis. Separation can be achieved by either normal-phase (NP) HPLC or reverse-phase (RP) HPLC. For example, NP-HPLC effectively separates glycerophospholipids on the basis of headgroup polarity, whereas RP-HPLC effectively separates fatty acids such as eicosanoids on the basis of chain length, degree of unsaturation and substitution. For global, untargeted lipidomic studies it is common to use both RP and NP or Hydrophilic

Interaction Liquid Chromatrography (HILC) columns for increased coverage of lipidome. This can be further improved with ultra-performance (UPLC) columns (2.1 mm ID & sub-2 micrometer particle size) or columns based on solid- core technology (2.1 mm ID & 2.6 μm particle size). The UHPLC columns allow high resolution separation of complex lipids with increased peak capacity and sensitivity, whereas solid- core columns achieve same at shorter times on normal HPLC pumps. Chromatographic (HPLC/UHPLC) separation of lipids may either be performed offline or online where the eluate is integrated with the ionization source of a mass spectrometer.

Lipid Detection

The progress of modern lipidomics has been greatly accelerated by the development of spectrometric methods in general and soft ionization techniques for mass spectrometry such as electrospray ionization (ESI) and matrix-assisted laser desorption/ionization (MALDI) in particular. "Soft" ionization does not cause extensive fragmentation, so that comprehensive detection of an entire range of lipids within a complex mixture can be correlated to experimental conditions or disease state. In addition to ESI and MALDI, the technique of atmospheric pressure chemical ionization (APCI) has become increasingly popular for the analysis of nonpolar lipids.

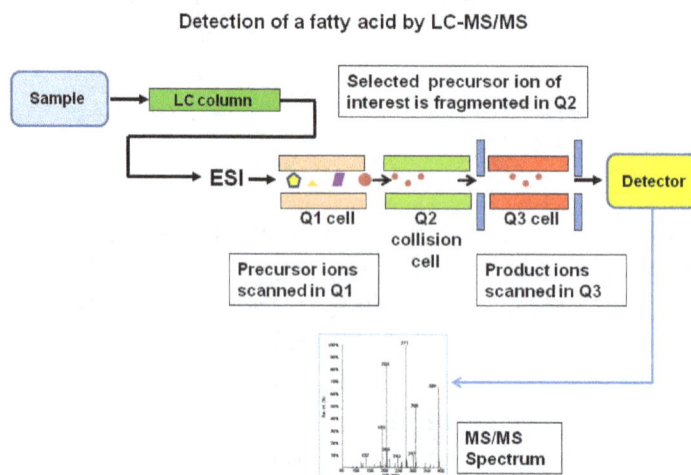

Schema showing detection of a fatty acid by LC-MS/MS using a linear ion-trap instrument and an electrospray (ESI) ion source.

ESI MS

ESI-MS was initially developed by Fenn and colleagues for analysis of biomolecules. It depends on the formation of gaseous ions from polar, thermally labile and mostly non-volatile molecules and thus is completely suitable for a variety of lipids. It is a soft-ionization method that rarely disrupts the chemical nature of the analyte prior to mass analysis. Various ESI-MS methods have been developed for analysis of different classes, subclasses, and individual lipid species from biological extracts. Comprehensive reviews of the methods and their application have recently been published. The major advantages of ESI-MS are high accuracy, sensitivity, reproducibility, and the applicability of the technique to complex solutions without prior derivatization. Han and coworkers have developed a method known as"shotgun lipidomics" which involves direct infusion of a crude lipid extract into an ESI source optimized for intrasource separation of lipids based on their intrinsic electrical properties.

MALDI MS

MALDI mass spectrometry is a laser-based soft-ionization method often used for analysis of large proteins, but has been used successfully for lipids. The lipid is mixed with a matrix, such as 2,5-dihydroxybenzoic acid, and applied to a sample holder as a small spot. A laser is fired at the spot, and the matrix absorbs the energy, which is then transferred to the analyte, resulting in ionization of the molecule. MALDI-Time-of-flight (MALDI-TOF) MS has become a very promising approach for lipidomics studies, particularly for the imaging of lipids from tissue slides.

APCI MS

The source for APCI is similar to ESI except that ions are formed by the interaction of the heated analyte solvent with a corona discharge needle set at a high electrical potential. Primary ions are formed immediately surrounding the needle, and these interact with the solvent to form secondary ions that ultimately ionize the sample. APCI is particularly useful for the analysis of nonpolar lipids such as triacylglycerols, sterols, and fatty acid esters.

Imaging Techniques

Recent developments in MALDI methods have enabled direct detection of lipids in-situ. Abundant lipid-related ions are produced from the direct analysis of thin tissue slices when sequential spectra are acquired across a tissue surface that has been coated with a MALDI matrix. Collisional activation of the molecular ions can be used to determine the lipid family and often structurally define the molecular species. This technique enables detection of phospholipids, sphingolipids and glycerolipids in tissues such as heart, kidney and brain. Furthermore, distribution of many different lipid molecular species often define anatomical regions within these tissues.

Lipidomic Profiling

Quantitative lipid profiles (lipidomes) of yeast Saccharomyces cerevisiae grown in different temperatures

Lipid profiling is a targeted metabolomics platform that provides a comprehensive analysis of lipid species within a cell or tissue. Profiling based on electrospray ionization tandem mass spectrometry (ESI-MS/MS) is capable of providing quantitative data and is adaptable to high throughput analyses. The powerful approach of transgenics, namely deletion and/or overexpression of a gene product coupled with lipidomics, can give valuable insights into the role of biochemical pathways. Lipid profiling techniques have also been applied to plants and microorganisms such as yeast. A combination of quantitative lipidomic data in conjunction with the corresponding transcriptional data (using gene-array methods) and proteomic data (using tandem MS) enables a systems biology approach to a more in-depth understanding of the metabolic or signaling pathways of interest.

Informatics

A major challenge for lipidomics, in particular for MS-based approaches, lies in the computational and bioinformatic demands of handling the large amount of data that arise at various stages along the chain of information acquisition and processing. Chromatographic and MS data collection requires substantial efforts in spectral alignment and statistical evaluation of fluctuations in signal intensities. Such variations have a multitude of origins, including biological variations, sample handling and analytical accuracy. As a consequence several replicates are normally required for reliable determination of lipid levels in complex mixtures. Within the last few years, a number of software packages have been developed by various companies and research groups to analyze data generated by MS profiling of metabolites, including lipids. The data processing for differential profiling usually proceed through several stages, including input file manipulation, spectral filtering, peak detection, chromatographic alignment, normalization, visualization, and data export. An example of metabolic profiling software is the freely-available Java-based Mzmine application. Some software packages such as Markerview include multivariate statistical analysis (for example, principal component analysis) and these will be helpful for the identification of correlations in lipid metabolites that are associated with a physiological phenotype, in particular for the development of lipid-based biomarkers.Another objective of the information technology side of lipidomics involves the construction of metabolic maps from data on lipid structures and lipid-related protein and genes. Some of these lipid pathways are extremely complex, for example the mammalian glycosphingolipid pathway. The establishment of searchable and interactive databases of lipids and lipid-related genes/proteins is also an extremely important resource as a reference for the lipidomics community. Integration of these databases with MS and other experimental data, as well as with metabolic networks offers an opportunity to devise therapeutic strategies to prevent or reverse these pathological states involving dysfunction of lipid-related processes.

Protein-Lipid Interaction

Protein–lipid interaction is the influence of membrane proteins on the lipid physical state or vice versa.

The questions which are relevant to understanding of the structure and function of the membrane are: 1) Do intrinsic membrane proteins bind tightly to lipids, and what is the nature of the layer of lipids adjacent to the protein? 2) Do membrane proteins have long-range effects on the order or dynamics of membrane lipids? 3) How do the lipids influence the structure and/or function of membrane proteins? 4) How do peripheral membrane proteins which bind to the layer surface interact with lipids and influence their behavior?

Binding of Lipids to Intrinsic Membrane Proteins in the Bilayer

A large research effort involves approaches to know whether proteins have binding sites which are specific for particular lipids and whether the protein–lipid complexes can be considered to be long-lived, on the order of the time required for the turnover a typical enzyme, that is 10^- sec. This is now known through the use of H-NMR, ESR, and fluorescent methods.

There are two approaches used to measure the relative affinity of lipids binding to specific membrane proteins. These involve the use of lipid analogues in reconstituted phospholipid vesicles containing the protein of interest: 1) Spin-labeled phospholipids are motionally restricted when they are adjacent to membrane proteins. The result is a component in the ESR spectrum which is broadened. The experimental spectrum can be analyzed as the sum of the two components, a rapidly tumbling species in the "bulk" lipid phase with a sharp spectrum, and a motionally restricted component adjacent to the protein. Membrane protein denaturation causes further broadening of ESR spin label spectrum and throws more light on membrane lipid-proteins interactions 2) Spin-labeled and brominated lipid derivatives are able to quench the intrinsic tryptophan fluorescence from membrane proteins. The efficiency of quenching depends on the distance between the lipid derivative and the fluorescent tryptophans.

Perturbations of the Lipid Bilayer Due to the Presence of Lateral Membrane Proteins

Most H-NMR experiments with deuterated phospholipids demonstrate that the presence of proteins has little effect on either the order parameter of the lipids in the bilayer or the lipid dynamics, as measured by relaxation times. The overall view resulting from NMR experiments is 1) that the exchange rate between boundary and free lipids is rapid, (10 sec⁻), 2) that the order parameters of the bound lipid are barely affected by being adjacent to proteins, 3) that the dynamics of the acyl chain reorientations are slowed only slightly in the frequency range of 10 sec⁻, and 4) that the orientation and the dynamics of the polar headgroups are similarly unaffected in any substantial manner by being adjacent to transmembrane proteins. 13C-NMR spectrum also gives information on specific lipid-protein interactions of biomembranes

Recent results using non labeled optical methods such as Dual Polarisation Interferometry which measure the birefringence(or order) within lipid bilayers have been used to show how peptide and protein interactions can influence bilayer order, specifically demonstrating the real time association to bilayer and critical peptide concentration after which the peptides penetrate and disrupt the bilayer order.

Backbone and Solid Chain Dynamics of Membrane Proteins

Solid-state NMR techniques have the potential to yield detailed information about the dynamics of individual amino acid residues within a membrane protein. However, the techniques can require large amounts (100–200 mg) of isotopically labeled proteins and are most informative when applied to small proteins where spectroscopic assignments are possible.

Binding of Peripheral Membrane Proteins to the Lipid Bilayer

Many peripheral membrane proteins bind to the membrane primarily through interactions with integral membrane proteins. But there is a diverse group of proteins which interact directly with the surface of the lipid bilayer. Some, such as myelin basic protein, and spectrin have mainly structural roles. A number of water-soluble proteins can bind to the bilayer surface transiently or under specific conditions.

Misfolding processes, typically exposing hydrophobic regions of proteins, often are associated with binding to lipid membranes and subsequent aggregation, for example, during neurodegenerative disorders, neuronal stress and apoptosis.

Phenolic Lipids

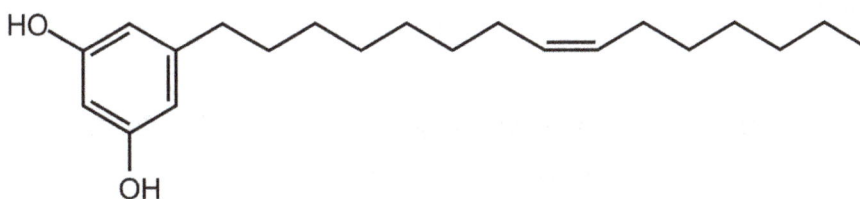

Bilobol, an alkylresorcinol found in Ginkgo biloba

Phenolic lipids are a class of natural products composed of long aliphatic chains and phenolic rings. Phenolic lipids occur in plants, fungi and bacteria.

Types

- Alkylcatechols
- Alkylphenols (nonylphenol, cardanol)
- Alkylresorcinols
- Anacardic acids

Biological Activity

Due to their strong amphiphilic character, the phenolic lipids can incorporate into erythrocytes and liposomal membranes. The ability of these compounds to inhibit bacterial, fungal, protozoan and parasite growth seems to depend on their interaction with proteins and/or on their membrane-disturbing properties.

Biological Role

The phenolic lipid synthesis by type III polyketide synthases is essential for cyst formation in Azotobacter vinelandii.

Types of Phenolic Lipids

Alkylphenols

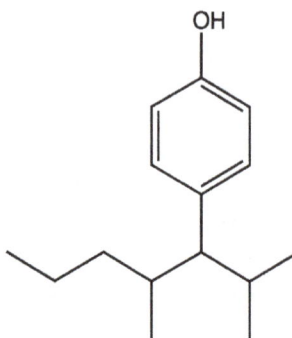

Chemical structure of the alkylphenol nonylphenol

Alkylphenols are a family of organic compounds obtained by the alkylation of phenols. The term is usually reserved for commercially important propylphenol, butylphenol, amylphenol, heptylphenol, octylphenol, nonylphenol, dodecylphenol and related "long chain alkylphenols" (LCAPs). Methylphenols and ethylphenols are also alkylphenols, but they are more commonly referred to by their specific names, cresols and xylenols.

Production

The long-chain alkylphenols are prepared by alkylation of phenol with alkenes:

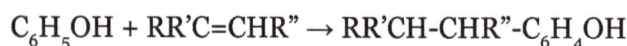

$$C_6H_5OH + RR'C=CHR'' \rightarrow RR'CH\text{-}CHR''\text{-}C_6H_4OH$$

In this way, about 500M kg/y are produced.

Environmental Controversy Over Nonylphenols

Alkylphenols are xenoestrogens. The European Union has implemented sales and use restrictions on certain applications in which nonylphenols are used because of their alleged "toxicity, persistence, and the liability to bioaccumulate" but the United States EPA has taken a slower approach to make sure that action is based on "sound science".

Uses of Long-Chain Alkylphenols

The long-chain alkylphenols are used extensively as precursors to the detergents, as additives for fuels and lubricants, polymers, and as components in phenolic resins. These compounds are also used as building block chemicals that are also used in making fragrances, thermoplastic elastomers, antioxidants, oil field chemicals and fire retardant materials. Through the downstream use in making alkylphenolic resins, alkylphenols are also found in tires, adhesives, coatings, carbonless copypaper and high performance rubber products. They have been used in industry for over 40 years.

These xenobiotic compounds are known to be weak endocrine disruptors.

Alkylresorcinol

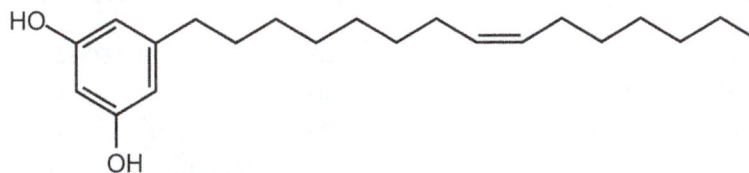

Bilobol (5-[(Z)-pentadec-8-enylo]resorcinol)

Alkylresorcinols, also known as resorcinolic lipids, are phenolic lipids composed of long aliphatic chains and resorcinol-type phenolic rings.

Natural Sources of Alkylresorcinols

Alkylresorcinols are relatively rare in nature, with the main known sources being wheat, rye, barley, triticale (cereal grasses), the fruit of Ginkgo biloba Bilobol, a mollusc and some species of bacteria.

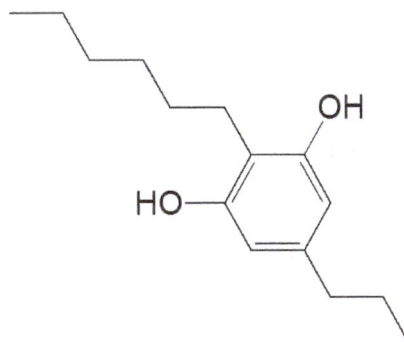

DB-2073 (2-n-hexyl-5-n-propylresorcinol)

DB-2073 is an antibiotic isolated from the broth culture of Pseudomonas sp. They are also the main constituents of the outer shell of the cyst of Azotobacter.

Occurrence in Cereals

Alkylresorcinols are present in high amounts in the bran layer (e.g. pericarp, testa and aleurone layers) of wheat and rye (0.1-0.3 % of dry weight). 5-Alkylresorcinols can also be found in rice, though not in the edible parts of the rice plant.

They are only present in very low amounts in the endosperm (the part of cereal grain that is used to make white flour), which means that alkylresorcinols can be used as 'biomarkers' for people who eat foods containing wholegrain wheat and rye, rather than cereal products based on white flour.

Alkylresorcinols were thought to have anti-nutritive properties (e.g. decreasing growth of pigs and chickens fed rye), but this theory has been discredited, and a number of animal studies have demonstrated that they have no obvious negative effect on animals or humans.

Biomarkers of a Whole Grain Diet

Increasing evidence from human intervention trials suggests that they are the most promising biomarker of whole grain wheat and rye intake. Alkylresorcinol metabolites, 3,5-dihydroxybenzoic acid (DHBA) and 3,5-dihydroxyphenylpropionoic acid (DHPPA) were first identified in urine and can be quantified in urine and plasma, and may be an alternative, equivalent biomarker of whole grain wheat intake.

The average intake of alkylresorcinols in the UK is around 11 mg/person/day, and in Sweden is around 20 mg/person/day. This varies widely depending on whether people normally consume wholegrain/wholemeal/brown bread, which is high in alkylresorcinols (300-1000 µg/g), or white wheat bread, which has very low concentrations of alkylresorcinols (<50 µg/g).

Possible Biological Activities

In vitro studies have shown that alkylresorcinols may prevent cells turning cancerous, but that they do not have any effect on cells that are already cancerous. Alkylresorcinols also increase gamma-tocopherol levels in rats when fed in high amounts (0.2% of total diet and above.

The alkylresorcinols in Grevillea banksii and Grevillea 'Robyn Gordon' are responsible for contact dermatitis.

Trivial names of Some Resorcinolic Lipids

- olivetol
- ardisinol I
- adipostatin B
- hydrobilobol
- irisresorcinol
- R-leprosol
- xenognosin

- persoonol
- ardisinol II
- bilobol
- cardol
- panosialin
- α-leprosol

- grevillol
- adipostatin A
- hexylresorcinol
- rucinol
- stemphol
- merulinic acid

Derivatives

Sorgoleone is a hydrophobic root exudate of Sorghum bicolor.

Anacardic Acids

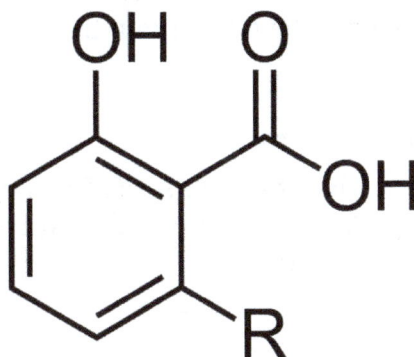

General structure of anacardic acids. R is an alkyl chain of variable length, which may be saturated or unsaturated.

Anacardic acids are phenolic lipids, chemical compounds found in the shell of the cashew nut (Anacardium occidentale). An acid form of urushiol, they also cause an allergic skin rash on contact, known as urushiol-induced contact dermatitis. Anacardic acid is a yellow liquid. It is partially miscible with ethanol and ether, but nearly immiscible with water. Chemically, anacardic acid is a mixture of several closely related organic compounds. Each consists of a salicylic acid substituted with an alkyl chain that has 15 or 17 carbon atoms. The alkyl group may be saturated or unsaturated; anacardic acid is a mixture of saturated and unsaturated molecules. The exact mixture depends on the species of the plant. The 15 carbon unsaturated side chain compound found in the cashew plant is lethal to Gram positive bacteria.

Folk use for tooth abscesses, it is also active against acne, some insects, tuberculosis, and MRSA. It is primarily found in foods such as cashew nuts, cashew apples, and cashew nutshell oil, but also in mangos and Pelargonium geraniums.

Experimental Antibacterial Properties

The side chain with three unsaturated bonds was the most active against Streptococcus mutans, the tooth decay bacterium, in test tube experiments. The number of unsaturated bonds was not material against Propionibacterium acnes, the acne bacterium. Eichbaum claims that a solution of one part anacardic acid to 200,000 parts water to as low as one part in 2,000,000 is lethal to Gram positive bacteria in 15 minutes in vitro. Somewhat higher ratios killed tubercle bacteria of tuberculosis in 30 minutes. Heating these anacardic acids converts them to the alcohols (cardanols) with reduced activity compared to the acids. Decarboxylation, such as through heating done in most commercial oil processing, results in compounds with significantly reduced activity. It is said that the people of the Gold Coast use cashew leaves and bark for a toothache.

Industrial Uses

Anacardic acid is the main component of cashew nutshell liquid (CNSL), and finds use in the chemical industry for the production of cardanol, which is used for resins, coatings, and frictional materials. Cardanol is used to make phenalkamines, which are used as curing agents for the durable epoxy coatings used on concrete floors.

History

The first chemical analysis of the oil of the cashew nut shell from the Anacardium occidentale was published in 1847. It was later found to be a mixture rather than one chemical, sometimes the plural anacardic acids is used.

Synergies

Anacardic acid is synergistic with anethole from the seed of anise (Umbelliferae) and linalool from green tea in vitro [Muroi & Kubo, p1782]. The totarol in the bark of Podocarpus trees is synergistic with anacardic acid in its bactericidal effects.

Other and Potential Uses

There is also a suspicion that inhibiting anacardic acids may arrest the growth of cancer tumors such as breast cancer. [Kubo et al., 1993]

Anacardic acid (2-hydroxy-6-alkylbenzoic acid) provides resistance to small pest insects (aphids and spider mites).

Anacardic acid kills methicillin-resistant Staphylococcus aureus (MRSA) cells more rapidly than totarol.

List of Anacardic Acids

- 6-pentadecyl salicylic acid (6-PDSA), a potent HAT inhibitor from cashew nut shell liquid, and sensitizer of cancer cells to ionizing radiation.

Nucleic Acid

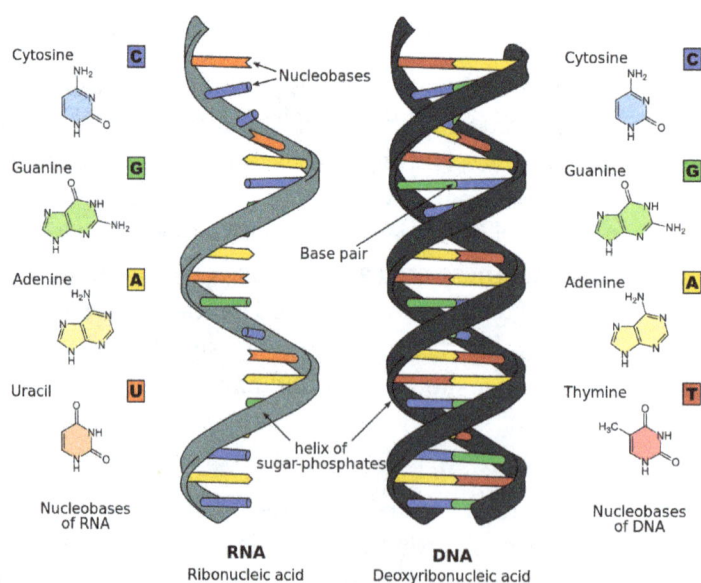

A comparison of the two principal nucleic acids: RNA (left) and DNA (right), showing the helices and nucleobases each employs.

Nucleic acids are biopolymers, or large biomolecules, essential for all known forms of life. Nucleic acids, which include DNA (deoxyribonucleic acid) and RNA (ribonucleic acid), are made from monomers known as nucleotides. Each nucleotide has three components: a 5-carbon sugar, a phosphate group, and a nitrogenous base. If the sugar is deoxyribose, the polymer is DNA. If the sugar is ribose, the polymer is RNA. When all three components are combined, they form a nucleic acid. Nucleotides are also known as phosphate nucleotides.

Nucleic acids are among the most important biological macromolecules (others being amino acids/proteins, sugars/carbohydrates, and lipids/fats). They are found in abundance in all living things, where they function in encoding, transmitting and expressing genetic information—in other words, information is conveyed through the nucleic acid sequence, or the order of nucleotides within a DNA or RNA molecule. Strings of nucleotides strung together in a specific sequence are the mechanism for storing and transmitting hereditary, or genetic information via protein synthesis.

Nucleic acids were discovered by Friedrich Miescher in 1869. Experimental studies of nucleic acids constitute a major part of modern biological and medical research, and form a foundation for genome and forensic science, as well as the biotechnology and pharmaceutical industries.

Occurrence and Nomenclature

The term nucleic acid is the overall name for DNA and RNA, members of a family of biopolymers, and is synonymous with polynucleotide. Nucleic acids were named for their initial discovery within the nucleus, and for the presence of phosphate groups (related to phosphoric acid). Although first discovered within the nucleus of eukaryotic cells, nucleic acids are now known to be found in all life forms including within bacteria, archaea, mitochondria, chloroplasts, viruses, and viroids. (note: there is debate as to whether viruses are living or non-living). All living cells contain both DNA and RNA (except some cells such as mature red blood cells), while viruses contain either DNA or RNA, but usually not both. The basic component of biological nucleic acids is the nucleotide, each of which contains a pentose sugar (ribose or deoxyribose), a phosphate group, and a nucleobase. Nucleic acids are also generated within the laboratory, through the use of enzymes (DNA and RNA polymerases) and by solid-phase chemical synthesis. The chemical methods also enable the generation of altered nucleic acids that are not found in nature, for example peptide nucleic acids.

Molecular Composition and Size

Nucleic acids are generally very large molecules. Indeed, DNA molecules are probably the largest individual molecules known. Well-studied biological nucleic acid molecules range in size from 21 nucleotides (small interfering RNA) to large chromosomes (human chromosome 1 is a single molecule that contains 247 million base pairs).

In most cases, naturally occurring DNA molecules are double-stranded and RNA molecules are single-stranded. There are numerous exceptions, however—some viruses have genomes made of double-stranded RNA and other viruses have single-stranded DNA genomes, and, in some circumstances, nucleic acid structures with three or four strands can form.

Nucleic acids are linear polymers (chains) of nucleotides. Each nucleotide consists of three components: a purine or pyrimidine nucleobase (sometimes termed nitrogenous base or simply base), a pentose sugar, and a phosphate group. The substructure consisting of a nucleobase plus sugar is termed a nucleoside. Nucleic acid types differ in the structure of the sugar in their nucleotides– DNA contains 2'-deoxyribose while RNA contains ribose (where the only difference is the presence of a hydroxyl group). Also, the nucleobases found in the two nucleic acid types are different: adenine, cytosine, and guanine are found in both RNA and DNA, while thymine occurs in DNA and uracil occurs in RNA.

The sugars and phosphates in nucleic acids are connected to each other in an alternating chain (sugar-phosphate backbone) through phosphodiester linkages. In conventional nomenclature, the carbons to which the phosphate groups attach are the 3'-end and the 5'-end carbons of the sugar. This gives nucleic acids directionality, and the ends of nucleic acid molecules are referred to as 5'-end and 3'-end. The nucleobases are joined to the sugars via an N-glycosidic linkage involving a nucleobase ring nitrogen (N-1 for pyrimidines and N-9 for purines) and the 1' carbon of the pentose sugar ring.

Non-standard nucleosides are also found in both RNA and DNA and usually arise from modification of the standard nucleosides within the DNA molecule or the primary (initial) RNA transcript. Transfer RNA (tRNA) molecules contain a particularly large number of modified nucleosides.

Topology

Double-stranded nucleic acids are made up of complementary sequences, in which extensive Watson-Crick base pairing results in a highly repeated and quite uniform double-helical three-dimensional structure. In contrast, single-stranded RNA and DNA molecules are not constrained to a regular double helix, and can adopt highly complex three-dimensional structures that are based on short stretches of intramolecular base-paired sequences that include both Watson-Crick and noncanonical base pairs, as well as a wide range of complex tertiary interactions.

Nucleic acid molecules are usually unbranched, and may occur as linear and circular molecules. For example, bacterial chromosomes, plasmids, mitochondrial DNA, and chloroplast DNA are usually circular double-stranded DNA molecules, while chromosomes of the eukaryotic nucleus are usually linear double-stranded DNA molecules. Most RNA molecules are linear, single-stranded molecules, but both circular and branched molecules can result from RNA splicing reactions.

Nucleic Acid Sequences

One DNA or RNA molecule differs from another primarily in the sequence of nucleotides. Nucleotide sequences are of great importance in biology since they carry the ultimate instructions that encode all biological molecules, molecular assemblies, subcellular and cellular structures, organs, and organisms, and directly enable cognition, memory, and behavior. Enormous efforts have gone into the development of experimental methods to determine the nucleotide sequence of biological DNA and RNA molecules, and today hundreds of millions of nucleotides are sequenced daily at genome centers and smaller laboratories worldwide. In addition to maintaining the GenBank nucleic acid sequence database, the National Center for Biotechnology Information (NCBI, http://www.ncbi.nlm.nih.gov) provides analysis and retrieval resources for the data in GenBank and other biological data made available through the NCBI Web site

Types of Nucleic Acids

Deoxyribonucleic Acid

Deoxyribonucleic acid (DNA) is a nucleic acid containing the genetic instructions used in the development and functioning of all known living organisms. The DNA segments carrying this genetic information are called genes. Likewise, other DNA sequences have structural purposes, or are involved in regulating the use of this genetic information. Along with RNA and proteins, DNA is one of the three major macromolecules that are essential for all known forms of life. DNA consists of two long polymers of simple units called nucleotides, with backbones made of sugars and phosphate groups joined by ester bonds. These two strands run in opposite directions to each other and are, therefore, anti-parallel. Attached to each sugar is one of four types of molecules called nucleobases (informally, bases). It is the sequence of these four nucleobases along the backbone that encodes information. This information is read using the genetic code, which specifies the sequence of the amino acids within proteins. The code is read by copying stretches of DNA into the related nucleic acid RNA in a process called transcription. Within cells DNA is organized into long structures called chromosomes. During cell division these chromosomes are duplicated in the process of DNA replication, providing each cell its own complete set of chromosomes. Eukaryotic organisms (animals, plants, fungi, and protists) store most of their DNA inside the cell nucleus and some of their DNA in organelles, such as mitochondria or chloroplasts. In contrast, prokaryotes (bacteria and archaea) store their DNA only in the cytoplasm. Within the chromosomes, chromatin proteins such as histones compact and organize DNA. These compact structures guide the interactions between DNA and other proteins, helping control which parts of the DNA are transcribed.

Ribonucleic Acid

Ribonucleic acid (RNA) functions in converting genetic information from genes into the amino acid sequences of proteins. The three universal types of RNA include transfer RNA (tRNA), messenger RNA (mRNA), and ribosomal RNA (rRNA). Messenger RNA acts to carry genetic sequence information between DNA and ribosomes, directing protein synthesis. Ribosomal RNA is a major component of the ribosome, and catalyzes peptide bond formation. Transfer RNA serves as the carrier molecule for amino acids to be used in protein synthesis, and is responsible for decoding the mRNA. In addition, many other classes of RNA are now known.

Artificial Nucleic Acid Analogs

Artificial nucleic acid analogues have been designed and synthesized by chemists, and include peptide nucleic acid, morpholino- and locked nucleic acid, as well as glycol nucleic acid and threose nucleic acid. Each of these is distinguished from naturally occurring DNA or RNA by changes to the backbone of the molecule.

Types of Nucleic Acid

DNA

Deoxyribonucleic acid is a molecule that carries the genetic instructions used in the growth, development, functioning and reproduction of all known living organisms and many viruses. DNA and RNA are nucleic acids; alongside proteins and complex carbohydrates (polysaccharides), they are one of the three major types of macromolecule that are essential for all known forms of life. Most DNA molecules consist of two biopolymer strands coiled around each other to form a double helix.

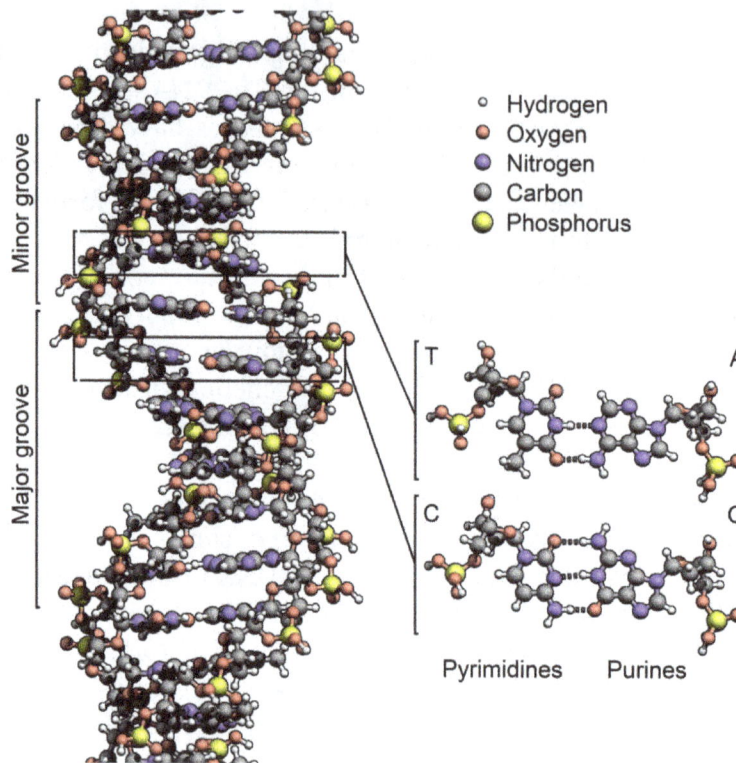

The structure of the DNA double helix. The atoms in the structure are colour-coded by element and the detailed structure of two base pairs are shown in the bottom right.

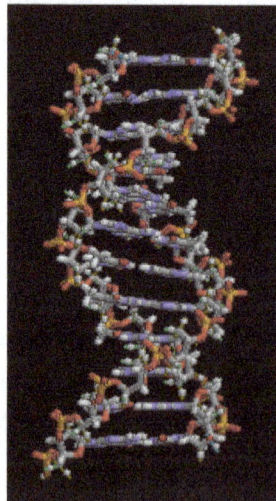

The structure of part of a DNA double helix

The two DNA strands are known as polynucleotides since they are composed of simpler units called nucleotides. Each nucleotide is composed of a nitrogen-containing nucleobase—either cytosine (C), guanine (G), adenine (A), or thymine (T)—as well as a sugar called deoxyribose and a phosphate group. The nucleotides are joined to one another in a chain by covalent bonds between the sugar of one nucleotide and the phosphate of the next, resulting in an alternating sugar-phosphate backbone. The nitrogenous bases of the two separate polynucleotide strands are bound together (according to base pairing rules (A with T, and C with G)) with hydrogen bonds to make double-stranded DNA. The total amount of related DNA base pairs on Earth is estimated at 5.0 x 10, and weighs 50 billion tonnes. In comparison, the total mass of the biosphere has been estimated to be as much as 4 TtC (trillion tons of carbon).

DNA stores biological information. The DNA backbone is resistant to cleavage, and both strands of the double-stranded structure store the same biological information. Biological information is replicated as the two strands are separated. A significant portion of DNA (more than 98% for humans) is non-coding, meaning that these sections do not serve as patterns for protein sequences.

The two strands of DNA run in opposite directions to each other and are therefore anti-parallel. Attached to each sugar is one of four types of nucleobases (informally, bases). It is the sequence of these four nucleobases along the backbone that encodes biological information. RNA strands are created using DNA strands as a template in a process called transcription. Under the genetic code, these RNA strands are translated to specify the sequence of amino acids within proteins in a process called translation.

Within eukaryotic cells, DNA is organized into long structures called chromosomes. During cell division these chromosomes are duplicated in the process of DNA replication, providing each cell its own complete set of chromosomes. Eukaryotic organisms (animals, plants, fungi, and protists) store most of their DNA inside the cell nucleus and some of their DNA in organelles, such as mitochondria or chloroplasts. In contrast, prokaryotes (bacteria and archaea) store their DNA only in the cytoplasm. Within the eukaryotic chromosomes, chromatin proteins such as histones compact and organize DNA. These compact structures guide the interactions between DNA and other proteins, helping control which parts of the DNA are transcribed.

DNA was first isolated by Friedrich Miescher in 1869. Its molecular structure was identified by James Watson and Francis Crick in 1953, whose model-building efforts were guided by X-ray diffraction data acquired by Rosalind Franklin. DNA is used by researchers as a molecular tool to explore physical laws and theories, such as the ergodic theorem and the theory of elasticity. The unique material properties of DNA have made it an attractive molecule for material scientists and engineers interested in micro- and nano-fabrication. Among notable advances in this field are DNA origami and DNA-based hybrid materials.

Properties

DNA is a long polymer made from repeating units called nucleotides. The structure of DNA is non-static, all species comprises two helical chains each coiled round the same axis, and each with a pitch of 34 ångströms (3.4 nanometres) and a radius of 10 ångströms (1.0 nanometre). According to another study, when measured in a particular solution, the DNA chain measured 22 to 26 ångströms wide (2.2 to 2.6 nanometres), and one nucleotide unit measured 3.3 Å (0.33 nm) long. Although each individual repeating unit is very small, DNA polymers can be very large molecules containing millions of nucleotides. For instance, the DNA in the largest human chromosome, chromosome number 1, consists of approximately 220 million base pairs and would be 85 mm long if straightened.

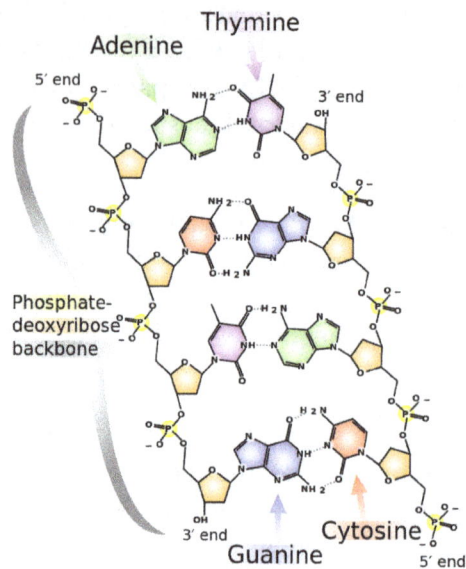

Chemical structure of DNA; hydrogen bonds shown as dotted lines

In living organisms DNA does not usually exist as a single molecule, but instead as a pair of molecules that are held tightly together. These two long strands entwine like vines, in the shape of a double helix. The nucleotide contains both a segment of the backbone of the molecule (which holds the chain together) and a nucleobase (which interacts with the other DNA strand in the helix). A nucleobase linked to a sugar is called a nucleoside and a base linked to a sugar and one or more phosphate groups is called a nucleotide. A polymer comprising multiple linked nucleotides (as in DNA) is called a polynucleotide.

The backbone of the DNA strand is made from alternating phosphate and sugar residues. The sugar in DNA is 2-deoxyribose, which is a pentose (five-carbon) sugar. The sugars are joined together by phosphate groups that form phosphodiester bonds between the third and fifth carbon atoms of adjacent sugar rings. These asymmetric bonds mean a strand of DNA has a direction. In a double helix the direction of the nucleotides in one strand is opposite to their direction in the other strand: the strands are antiparallel. The asymmetric ends of DNA strands are called the 5′ (five prime) and 3′ (three prime) ends, with the 5′ end having a terminal phosphate group and the 3′ end a terminal hydroxyl group. One major difference between DNA and RNA is the sugar, with the 2-deoxyribose in DNA being replaced by the alternative pentose sugar ribose in RNA.

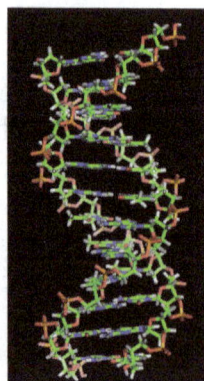

A section of DNA. The bases lie horizontally between the two spiraling strands. (animated version).

The DNA double helix is stabilized primarily by two forces: hydrogen bonds between nucleotides and base-stacking interactions among aromatic nucleobases. In the aqueous environment of the cell, the conjugated π bonds of nucleotide bases align perpendicular to the axis of the DNA molecule, minimizing their interaction with the solvation shell and therefore, the Gibbs free energy. The four bases found in DNA are adenine (abbreviated A), cytosine (C), guanine (G) and thymine (T). These four bases are attached to the sugar/phosphate to form the complete nucleotide, as shown for adenosine monophosphate. Adenine pairs with thymine and guanine pairs with cytosine. It was represented by A-T base pairs and G-C base pairs.

Nucleobase Classification

The nucleobases are classified into two types: the purines, A and G, being fused five- and six-membered heterocyclic compounds, and the pyrimidines, the six-membered rings C and T. A fifth pyrimidine nucleobase, uracil (U), usually takes the place of thymine in RNA and differs from thymine by lacking a methyl group on its ring. In addition to RNA and DNA a large number of artificial nucleic acid analogues have also been created to study the properties of nucleic acids, or for use in biotechnology.

Uracil is not usually found in DNA, occurring only as a breakdown product of cytosine. However, in a number of bacteriophages – Bacillus subtilis bacteriophages PBS1 and PBS2 and Yersinia bacteriophage piR1-37 – thymine has been replaced by uracil. Another phage - Staphylococcal phage S6 - has been identified with a genome where thymine has been replaced by uracil.

Base J (beta-d-glucopyranosyloxymethyluracil), a modified form of uracil, is also found in a number of organisms: the flagellates Diplonema and Euglena, and all the kinetoplastid genera. Biosynthesis of J occurs in two steps: in the first step a specific thymidine in DNA is converted into hydroxymethyldeoxyuridine; in the second HOMedU is glycosylated to form J. Proteins that bind specifically to this base have been identified. These proteins appear to be distant relatives of the Tet1 oncogene that is involved in the pathogenesis of acute myeloid leukemia. J appears to act as a termination signal for RNA polymerase II.

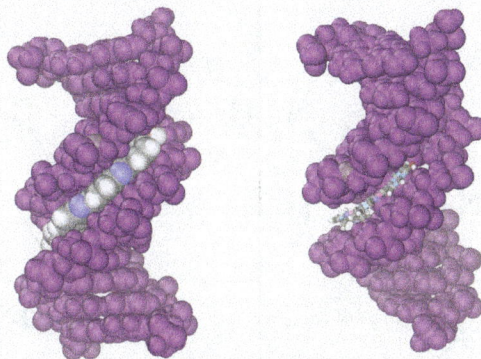

Major and minor grooves of DNA. Minor groove is a binding site for the dye Hoechst 33258.

Grooves

Twin helical strands form the DNA backbone. Another double helix may be found tracing the spaces, or grooves, between the strands. These voids are adjacent to the base pairs and may provide a binding site. As the strands are not symmetrically located with respect to each other, the grooves are unequally sized. One groove, the major groove, is 22 Å wide and the other, the minor groove, is 12 Å wide. The width of the major groove means that the edges of the bases are more accessible in the major groove than in the minor groove. As a result, proteins such as transcription factors that can bind to specific sequences in double-stranded DNA usually make contact with the sides of the bases exposed in the major groove. This situation varies in unusual conformations of DNA within the cell, but the major and minor grooves are always named to reflect the differences in size that would be seen if the DNA is twisted back into the ordinary B form.

Base Pairing

In a DNA double helix, each type of nucleobase on one strand bonds with just one type of nucleobase on the other strand. This is called complementary base pairing. Here, purines form hydrogen bonds to pyrimidines, with adenine bonding only to thymine in two hydrogen bonds, and cytosine bonding only to guanine in three hydrogen bonds. This arrangement of two nucleotides binding together across the double helix is called a base pair. As hydrogen bonds are not covalent, they can be broken and rejoined relatively easily. The two strands of DNA in a double helix can therefore be pulled apart like a zipper, either by a mechanical force or high temperature. As a result of this base pair complementarity, all the information in the double-stranded sequence of a DNA helix is duplicated on each strand, which is vital in DNA replication. Indeed, this reversible and specific interaction between complementary base pairs is critical for all the functions of DNA in living organisms.

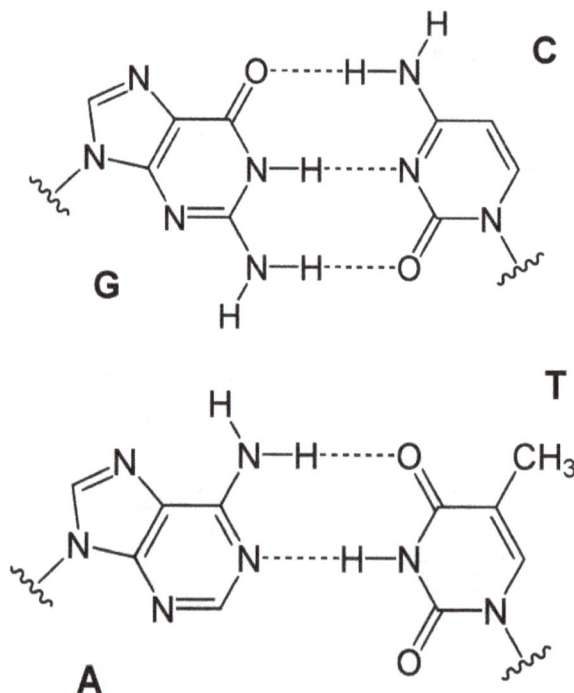

Top, a GC base pair with three hydrogen bonds. Bottom, an AT base pair with two hydrogen bonds. Non-covalent hydrogen bonds between the pairs are shown as dashed lines.

The two types of base pairs form different numbers of hydrogen bonds, AT forming two hydrogen bonds, and GC forming three hydrogen bonds. DNA with high GC-content is more stable than DNA with low GC-content.

As noted above, most DNA molecules are actually two polymer strands, bound together in a helical fashion by noncovalent bonds; this double stranded structure (dsDNA) is maintained largely by the intrastrand base stacking interactions, which are strongest for G,C stacks. The two strands can come apart – a process known as melting – to form two single-stranded DNA molecules (ssDNA) molecules. Melting occurs at high temperature, low salt and high pH (low pH also melts DNA, but since DNA is unstable due to acid depurination, low pH is rarely used).

The stability of the dsDNA form depends not only on the GC-content (% G,C basepairs) but also on sequence (since stacking is sequence specific) and also length (longer molecules are more stable). The stability can be measured in various ways; a common way is the "melting temperature", which is the temperature at which 50% of the ds molecules are converted to ss molecules; melting temperature is dependent on ionic strength and the concentration of DNA. As a result, it is both the percentage of GC base pairs and the overall length of a DNA double helix that determines the strength of the association between the two strands of DNA. Long DNA helices with a high GC-content have stronger-interacting strands, while short helices with high AT content have weaker-interacting strands. In biology, parts of the DNA double helix that need to separate easily, such as the TATAAT Pribnow box in some promoters, tend to have a high AT content, making the strands easier to pull apart.

In the laboratory, the strength of this interaction can be measured by finding the temperature necessary to break the hydrogen bonds, their melting temperature (also called T_m value). When all the base pairs in a DNA double helix melt, the strands separate and exist in solution as two entirely independent molecules. These single-stranded DNA molecules (ssDNA) have no single common shape, but some conformations are more stable than others.

Sense and Antisense

A DNA sequence is called "sense" if its sequence is the same as that of a messenger RNA copy that is translated into protein. The sequence on the opposite strand is called the "antisense" sequence. Both sense and antisense sequences can exist on different parts of the same strand of DNA (i.e. both strands can contain both sense and antisense sequences). In both prokaryotes and eukaryotes, antisense RNA sequences are produced, but the functions of these RNAs are not entirely clear. One proposal is that antisense RNAs are involved in regulating gene expression through RNA-RNA base pairing.

A few DNA sequences in prokaryotes and eukaryotes, and more in plasmids and viruses, blur the distinction between sense and antisense strands by having overlapping genes. In these cases, some DNA sequences do double duty, encoding one protein when read along one strand, and a second protein when read in the opposite direction along the other strand. In bacteria, this overlap may be involved in the regulation of gene transcription, while in viruses, overlapping genes increase the amount of information that can be encoded within the small viral genome.

Supercoiling

DNA can be twisted like a rope in a process called DNA supercoiling. With DNA in its "relaxed" state, a strand usually circles the axis of the double helix once every 10.4 base pairs, but if the DNA is twisted the strands become more tightly or more loosely wound. If the DNA is twisted in the direction of the helix, this is positive supercoiling, and the bases are held more tightly together. If they are twisted in the opposite direction, this is negative supercoiling, and the bases come apart more easily. In nature, most DNA has slight negative supercoiling that is introduced by enzymes called topoisomerases. These enzymes are also needed to relieve the twisting stresses introduced into DNA strands during processes such as transcription and DNA replication.

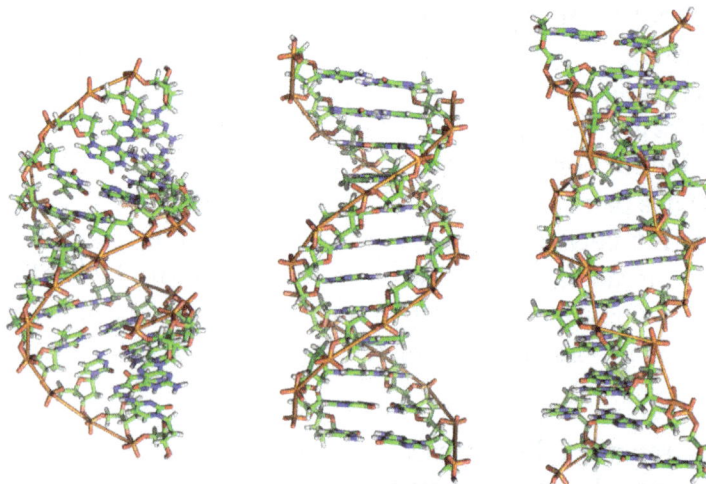

From left to right, the structures of A, B and Z DNA

Alternative DNA Structures

DNA exists in many possible conformations that include A-DNA, B-DNA, and Z-DNA forms, although, only B-DNA and Z-DNA have been directly observed in functional organisms. The conformation that DNA adopts depends on the hydration level, DNA sequence, the amount and direction of supercoiling, chemical modifications of the bases, the type and concentration of metal ions, as well as the presence of polyamines in solution.

The first published reports of A-DNA X-ray diffraction patterns—and also B-DNA—used analyses based on Patterson transforms that provided only a limited amount of structural information for oriented fibers of DNA. An alternative analysis was then proposed by Wilkins et al., in 1953, for the in vivo B-DNA X-ray diffraction/scattering patterns of highly hydrated DNA fibers in terms of squares of Bessel functions. In the same journal, James Watson and Francis Crick presented their molecular modeling analysis of the DNA X-ray diffraction patterns to suggest that the structure was a double-helix.

Although the "B-DNA form" is most common under the conditions found in cells, it is not a well-defined conformation but a family of related DNA conformations that occur at the high hydration levels present in living cells. Their corresponding X-ray diffraction and scattering patterns are characteristic of molecular paracrystals with a significant degree of disorder.

Compared to B-DNA, the A-DNA form is a wider right-handed spiral, with a shallow, wide minor groove and a narrower, deeper major groove. The A form occurs under non-physiological conditions in partially dehydrated samples of DNA, while in the cell it may be produced in hybrid pairings of DNA and RNA strands, as well as in enzyme-DNA complexes. Segments of DNA where the bases have been chemically modified by methylation may undergo a larger change in conformation and adopt the Z form. Here, the strands turn about the helical axis in a left-handed spiral, the opposite of the more common B form. These unusual structures can be recognized by specific Z-DNA binding proteins and may be involved in the regulation of transcription.

Alternative DNA Chemistry

For a number of years exobiologists have proposed the existence of a shadow biosphere, a postulated microbial biosphere of Earth that uses radically different biochemical and molecular processes than currently known life. One of the proposals was the existence of lifeforms that use arsenic instead of phosphorus in DNA. A report in 2010 of the possibility in the bacterium GFAJ-1, was announced, though the research was disputed, and evidence suggests the bacterium actively prevents the incorporation of arsenic into the DNA backbone and other biomolecules.

Quadruplex Structures

At the ends of the linear chromosomes are specialized regions of DNA called telomeres. The main function of these regions is to allow the cell to replicate chromosome ends using the enzyme telomerase, as the enzymes that normally replicate DNA cannot copy the extreme 3' ends of chromosomes. These specialized chromosome caps also help protect the DNA ends, and stop the DNA repair systems in the cell from treating them as damage to be corrected. In human cells, telomeres are usually lengths of single-stranded DNA containing several thousand repeats of a simple TTAGGG sequence.

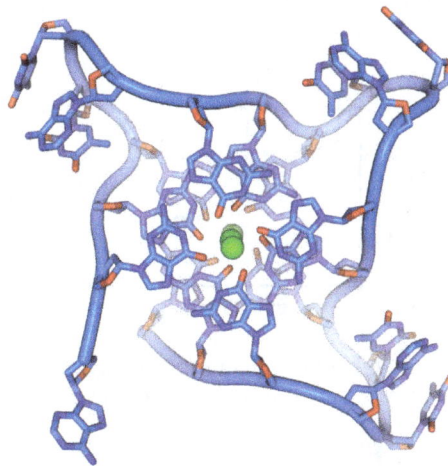

DNA quadruplex formed by telomere repeats.
The looped conformation of the DNA backbone is very different from the typical DNA helix.

These guanine-rich sequences may stabilize chromosome ends by forming structures of stacked sets of four-base units, rather than the usual base pairs found in other DNA molecules. Here, four guanine bases form a flat plate and these flat four-base units then stack on top of each other, to form a stable G-quadruplex structure. These structures are stabilized by hydrogen bonding between the edges of the bases and chelation of a metal ion in the centre of each four-base unit. Other structures can also be formed, with the central set of four bases coming from either a single strand folded around the bases, or several different parallel strands, each contributing one base to the central structure.

In addition to these stacked structures, telomeres also form large loop structures called telomere loops, or T-loops. Here, the single-stranded DNA curls around in a long circle stabilized by telomere-binding proteins. At the very end of the T-loop, the single-stranded telomere DNA is held onto a region of double-stranded DNA by the telomere strand disrupting the double-helical DNA and base pairing to one of the two strands. This triple-stranded structure is called a displacement loop or D-loop.

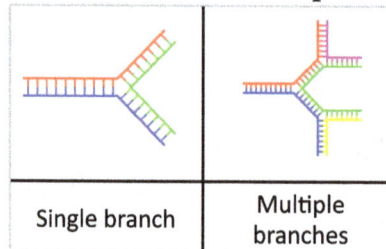

Branched DNA can form networks containing multiple branches.

Branched DNA

In DNA, fraying occurs when non-complementary regions exist at the end of an otherwise complementary double-strand of DNA. However, branched DNA can occur if a third strand of DNA is introduced and contains adjoining regions able to hybridize with the frayed regions of the pre-existing double-strand. Although the simplest example of branched DNA involves only three strands of DNA, complexes involving additional strands and multiple branches are also possible. Branched DNA can be used in nanotechnology to construct geometric shapes.

Chemical Modifications and Altered DNA Packaging

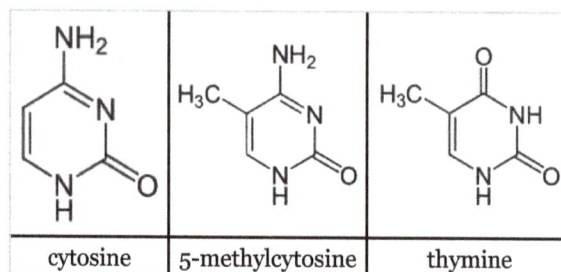

Structure of cytosine with and without the 5-methyl group. Deamination converts 5-methylcytosine into thymine.

Base Modifications and DNA Packaging

The expression of genes is influenced by how the DNA is packaged in chromosomes, in a structure called chromatin. Base modifications can be involved in packaging, with regions that have low or no gene expression usually containing high levels of methylation of cytosine bases. DNA packaging and its influence on gene expression can also occur by covalent modifications of the histone protein core around which DNA is wrapped in the chromatin structure or else by remodeling carried out by chromatin remodeling complexes. There is, further, crosstalk between DNA methylation and histone modification, so they can coordinately affect chromatin and gene expression.

For one example, cytosine methylation, produces 5-methylcytosine, which is important for X-chromosome inactivation. The average level of methylation varies between organisms – the worm Caenorhabditis elegans lacks cytosine methylation, while vertebrates have higher levels, with up to 1% of their DNA containing 5-methylcytosine. Despite the importance of 5-methylcytosine, it can deaminate to leave a thymine base, so methylated cytosines are particularly prone to mutations. Other base modifications include adenine methylation in bacteria, the presence of 5-hydroxymethylcytosine in the brain, and the glycosylation of uracil to produce the "J-base" in kinetoplastids.

Damage

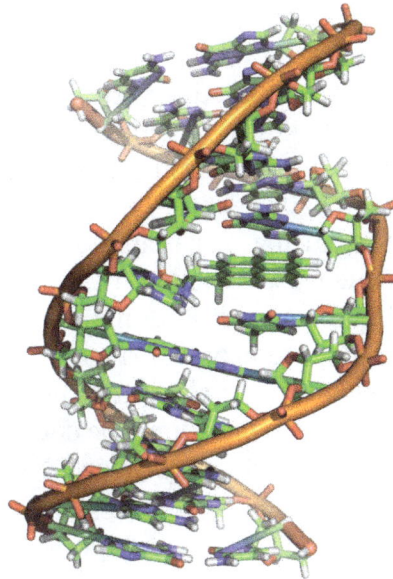

A covalent adduct between a metabolically activated form of benzo[a]pyrene, the major mutagen in tobacco smoke, and DNA

DNA can be damaged by many sorts of mutagens, which change the DNA sequence. Mutagens include oxidizing agents, alkylating agents and also high-energy electromagnetic radiation such as ultraviolet light and X-rays. The type of DNA damage produced depends on the type of mutagen. For example, UV light can damage DNA by producing thymine dimers, which are cross-links between pyrimidine bases. On the other hand, oxidants such as free radicals or hydrogen peroxide produce multiple forms of damage, including base modifications, particularly of guanosine, and double-strand breaks. A typical human cell contains about 150,000 bases that have suffered oxidative damage. Of these oxidative lesions, the most dangerous are double-strand breaks, as these are difficult to repair and can produce point mutations, insertions and deletions from the DNA sequence, as well as chromosomal translocations. These mutations can cause cancer. Because of inherent limitations in the DNA repair mechanisms, if humans lived long enough, they would all eventually develop cancer. DNA damages that are naturally occurring, due to normal cellular processes that produce reactive oxygen species, the hydrolytic activities of cellular water, etc., also occur frequently. Although most of these damages are repaired, in any cell some DNA damage may remain despite the action of repair processes. These remaining DNA damages accumulate with age in mammalian postmitotic tissues. This accumulation appears to be an important underlying cause of aging.

Many mutagens fit into the space between two adjacent base pairs, this is called intercalation. Most intercalators are aromatic and planar molecules; examples include ethidium bromide, acridines, daunomycin, and doxorubicin. For an intercalator to fit between base pairs, the bases must separate, distorting the DNA strands by unwinding of the double helix. This inhibits both transcription and DNA replication, causing toxicity and mutations. As a result, DNA intercalators may be carcinogens, and in the case of thalidomide, a teratogen. Others such as benzo[a]pyrene diol epoxide and aflatoxin form DNA adducts that induce errors in replication. Nevertheless, due to their ability to inhibit DNA transcription and replication, other similar toxins are also used in chemotherapy to inhibit rapidly growing cancer cells.

Biological Functions

DNA usually occurs as linear chromosomes in eukaryotes, and circular chromosomes in prokaryotes. The set of chromosomes in a cell makes up its genome; the human genome has approximately 3 billion base pairs of DNA arranged into 46 chromosomes. The information carried by DNA is held in the sequence of pieces of DNA called genes. Transmission of genetic information in genes is achieved via complementary base pairing. For example, in transcription, when a cell uses the information in a gene, the DNA sequence is copied into a complementary RNA sequence through the attraction between the DNA and the correct RNA nucleotides. Usually, this RNA copy is then used to make a matching protein sequence in a process called translation, which depends on the same interaction between RNA nucleotides. In alternative fashion, a cell may simply copy its genetic information in a process called DNA replication. The details of these functions are covered in other articles; here the focus is on the interactions between DNA and other molecules that mediate the function of the genome.

Genes and Genomes

Genomic DNA is tightly and orderly packed in the process called DNA condensation to fit the small available volumes of the cell. In eukaryotes, DNA is located in the cell nucleus, as well as small amounts in mitochondria and chloroplasts. In prokaryotes, the DNA is held within an irregularly shaped body in the cytoplasm called the nucleoid. The genetic information in a genome is held within genes, and the complete set of this information in an organism is called its genotype. A gene is a unit of heredity and is a region of DNA that influences a particular characteristic in an organism. Genes contain an open reading frame that can be transcribed, as well as regulatory sequences such as promoters and enhancers, which control the transcription of the open reading frame.

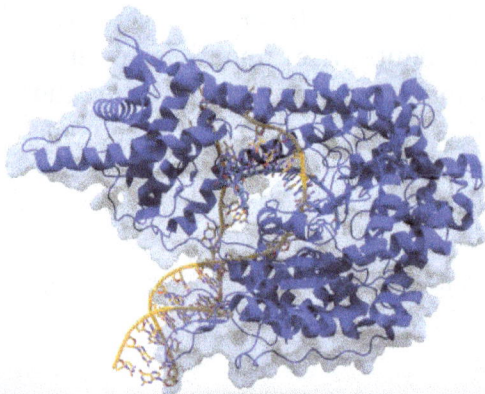

T7 RNA polymerase (blue) producing a mRNA (green) from a DNA template (orange).

In many species, only a small fraction of the total sequence of the genome encodes protein. For example, only about 1.5% of the human genome consists of protein-coding exons, with over 50% of human DNA consisting of non-coding repetitive sequences. The reasons for the presence of so much noncoding DNA in eukaryotic genomes and the extraordinary differences in genome size, or C-value, among species represent a long-standing puzzle known as the "C-value enigma". However, some DNA sequences that do not code protein may still encode functional non-coding RNA molecules, which are involved in the regulation of gene expression.

Some noncoding DNA sequences play structural roles in chromosomes. Telomeres and centromeres typically contain few genes, but are important for the function and stability of chromosomes. An abundant form of noncoding DNA in humans are pseudogenes, which are copies of genes that have been disabled by mutation. These sequences are usually just molecular fossils, although they can occasionally serve as raw genetic material for the creation of new genes through the process of gene duplication and divergence.

Transcription and Translation

A gene is a sequence of DNA that contains genetic information and can influence the phenotype of an organism. Within a gene, the sequence of bases along a DNA strand defines a messenger RNA sequence, which then defines one or more protein sequences. The relationship between the nucleotide sequences of genes and the amino-acid sequences of proteins is determined by the rules of translation, known collectively as the genetic code. The genetic code consists of three-letter 'words' called codons formed from a sequence of three nucleotides (e.g. ACT, CAG, TTT).

In transcription, the codons of a gene are copied into messenger RNA by RNA polymerase. This RNA copy is then decoded by a ribosome that reads the RNA sequence by base-pairing the messenger RNA to transfer RNA, which carries amino acids. Since there are 4 bases in 3-letter combinations, there are 64 possible codons (4 combinations). These encode the twenty standard amino acids, giving most amino acids more than one possible codon. There are also three 'stop' or 'nonsense' codons signifying the end of the coding region; these are the TAA, TGA, and TAG codons.

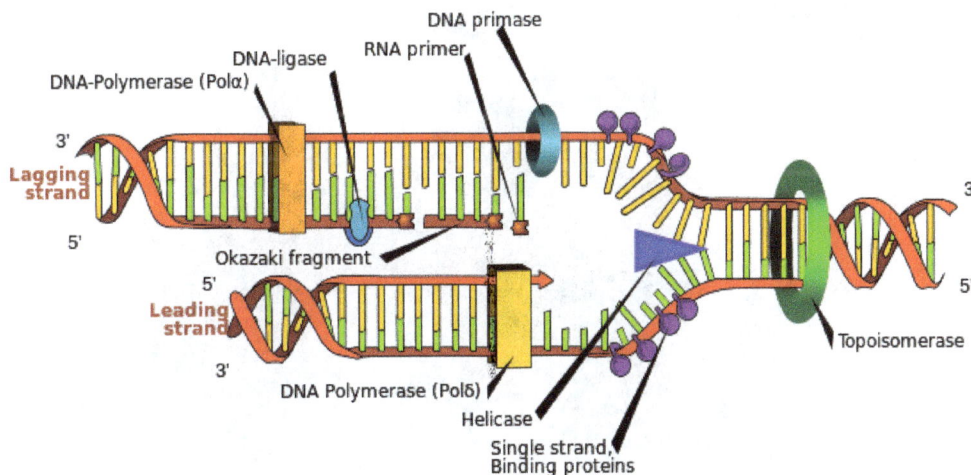

DNA replication. The double helix is unwound by a helicase and topoisomerase. Next, one DNA polymerase produces the leading strand copy. Another DNA polymerase binds to the lagging strand. This enzyme makes discontinuous segments (called Okazaki fragments) before DNA ligase joins them together.

Replication

Cell division is essential for an organism to grow, but, when a cell divides, it must replicate the DNA in its genome so that the two daughter cells have the same genetic information as their parent. The double-stranded structure of DNA provides a simple mechanism for DNA replication. Here, the two strands are separated and then each strand's complementary DNA sequence is recreated by an enzyme called DNA polymerase. This enzyme makes the complementary strand by finding the correct base through complementary base pairing, and bonding it onto the original strand. As DNA polymerases can only extend a DNA strand in a 5′ to 3′ direction, different mechanisms are used to copy the antiparallel strands of the double helix. In this way, the base on the old strand dictates which base appears on the new strand, and the cell ends up with a perfect copy of its DNA.

Extracellular Nucleic Acids

Naked extracellular DNA (eDNA), most of it released by cell death, is nearly ubiquitous in the environment. Its concentration in soil may be as high as 2 μg/L, and its concentration in natural aquatic environments may be as high at 88 μg/L. Various possible functions have been proposed for eDNA: it may be involved in horizontal gene transfer; it may provide nutrients; and it may act as a buffer to recruit or titrate ions or antibiotics. Extracellular DNA acts as a functional extracellular matrix component in the biofilms of a number of bacterial species. It may act as a recognition factor to regulate the attachment and dispersal of specific cell types in the biofilm; it may contribute to biofilm formation; and it may contribute to the biofilm's physical strength and resistance to biological stress.

Interactions with Proteins

All the functions of DNA depend on interactions with proteins. These protein interactions can be non-specific, or the protein can bind specifically to a single DNA sequence. Enzymes can also bind to DNA and of these, the polymerases that copy the DNA base sequence in transcription and DNA replication are particularly important.

DNA-Binding Proteins

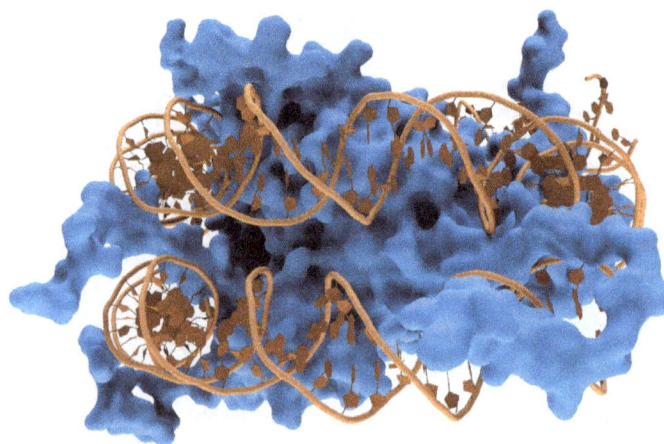

Interaction of DNA (shown in orange) with histones (shown in blue). These proteins' basic amino acids bind to the acidic phosphate groups on DNA.

Structural proteins that bind DNA are well-understood examples of non-specific DNA-protein interactions. Within chromosomes, DNA is held in complexes with structural proteins. These proteins organize the DNA into a compact structure called chromatin. In eukaryotes this structure involves DNA binding to a complex of small basic proteins called histones, while in prokaryotes multiple types of proteins are involved. The histones form a disk-shaped complex called a nucleosome, which contains two complete turns of double-stranded DNA wrapped around its surface. These non-specific interactions are formed through basic residues in the histones making ionic bonds to the acidic sugar-phosphate backbone of the DNA, and are therefore largely independent of the base sequence. Chemical modifications of these basic amino acid residues include methylation, phosphorylation and acetylation. These chemical changes alter the strength of the interaction between the DNA and the histones, making the DNA more or less accessible to transcription factors and changing the rate of transcription. Other non-specific DNA-binding proteins in chromatin include the high-mobility group proteins, which bind to bent or distorted DNA. These proteins are important in bending arrays of nucleosomes and arranging them into the larger structures that make up chromosomes.

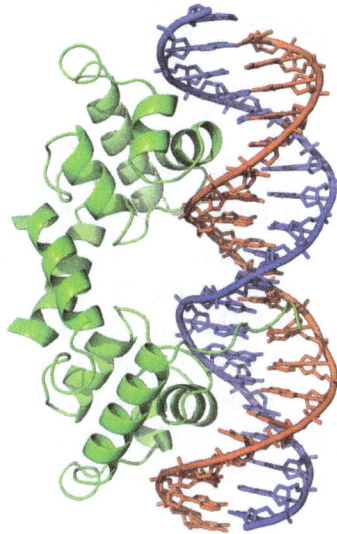

The lambda repressor helix-turn-helix transcription factor bound to its DNA target

A distinct group of DNA-binding proteins are the DNA-binding proteins that specifically bind single-stranded DNA. In humans, replication protein A is the best-understood member of this family and is used in processes where the double helix is separated, including DNA replication, recombination and DNA repair. These binding proteins seem to stabilize single-stranded DNA and protect it from forming stem-loops or being degraded by nucleases.

In contrast, other proteins have evolved to bind to particular DNA sequences. The most intensively studied of these are the various transcription factors, which are proteins that regulate transcription. Each transcription factor binds to one particular set of DNA sequences and activates or inhibits the transcription of genes that have these sequences close to their promoters. The transcription factors do this in two ways. Firstly, they can bind the RNA polymerase responsible for transcription, either directly or through other mediator proteins; this locates the polymerase at the promoter and allows it to begin transcription. Alternatively, transcription factors can bind enzymes that modify the histones at the promoter. This changes the accessibility of the DNA template to the polymerase.

As these DNA targets can occur throughout an organism's genome, changes in the activity of one type of transcription factor can affect thousands of genes. Consequently, these proteins are often the targets of the signal transduction processes that control responses to environmental changes or cellular differentiation and development. The specificity of these transcription factors' interactions with DNA come from the proteins making multiple contacts to the edges of the DNA bases, allowing them to "read" the DNA sequence. Most of these base-interactions are made in the major groove, where the bases are most accessible.

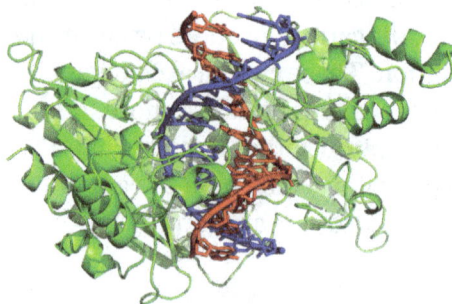

The restriction enzyme EcoRV (green) in a complex with its substrate DNA

DNA-Modifying Enzymes

Nucleases and Ligases

Nucleases are enzymes that cut DNA strands by catalyzing the hydrolysis of the phosphodiester bonds. Nucleases that hydrolyse nucleotides from the ends of DNA strands are called exonucleases, while endonucleases cut within strands. The most frequently used nucleases in molecular biology are the restriction endonucleases, which cut DNA at specific sequences. For instance, the EcoRV enzyme shown to the left recognizes the 6-base sequence 5′-GATATC-3′ and makes a cut at the vertical line. In nature, these enzymes protect bacteria against phage infection by digesting the phage DNA when it enters the bacterial cell, acting as part of the restriction modification system. In technology, these sequence-specific nucleases are used in molecular cloning and DNA fingerprinting.

Enzymes called DNA ligases can rejoin cut or broken DNA strands. Ligases are particularly important in lagging strand DNA replication, as they join together the short segments of DNA produced at the replication fork into a complete copy of the DNA template. They are also used in DNA repair and genetic recombination.

Topoisomerases and Helicases

Topoisomerases are enzymes with both nuclease and ligase activity. These proteins change the amount of supercoiling in DNA. Some of these enzymes work by cutting the DNA helix and allowing one section to rotate, thereby reducing its level of supercoiling; the enzyme then seals the DNA break. Other types of these enzymes are capable of cutting one DNA helix and then passing a second strand of DNA through this break, before rejoining the helix. Topoisomerases are required for many processes involving DNA, such as DNA replication and transcription.

Helicases are proteins that are a type of molecular motor. They use the chemical energy in nucleoside triphosphates, predominantly ATP, to break hydrogen bonds between bases and unwind the

DNA double helix into single strands. These enzymes are essential for most processes where enzymes need to access the DNA bases.

Polymerases

Polymerases are enzymes that synthesize polynucleotide chains from nucleoside triphosphates. The sequence of their products are created based on existing polynucleotide chains—which are called templates. These enzymes function by repeatedly adding a nucleotide to the 3′ hydroxyl group at the end of the growing polynucleotide chain. As a consequence, all polymerases work in a 5′ to 3′ direction. In the active site of these enzymes, the incoming nucleoside triphosphate base-pairs to the template: this allows polymerases to accurately synthesize the complementary strand of their template. Polymerases are classified according to the type of template that they use.

In DNA replication, DNA-dependent DNA polymerases make copies of DNA polynucleotide chains. In order to preserve biological information, it is essential that the sequence of bases in each copy are precisely complementary to the sequence of bases in the template strand. Many DNA polymerases have a proofreading activity. Here, the polymerase recognizes the occasional mistakes in the synthesis reaction by the lack of base pairing between the mismatched nucleotides. If a mismatch is detected, a 3′ to 5′ exonuclease activity is activated and the incorrect base removed. In most organisms, DNA polymerases function in a large complex called the replisome that contains multiple accessory subunits, such as the DNA clamp or helicases.

RNA-dependent DNA polymerases are a specialized class of polymerases that copy the sequence of an RNA strand into DNA. They include reverse transcriptase, which is a viral enzyme involved in the infection of cells by retroviruses, and telomerase, which is required for the replication of telomeres. Telomerase is an unusual polymerase because it contains its own RNA template as part of its structure.

Transcription is carried out by a DNA-dependent RNA polymerase that copies the sequence of a DNA strand into RNA. To begin transcribing a gene, the RNA polymerase binds to a sequence of DNA called a promoter and separates the DNA strands. It then copies the gene sequence into a messenger RNA transcript until it reaches a region of DNA called the terminator, where it halts and detaches from the DNA. As with human DNA-dependent DNA polymerases, RNA polymerase II, the enzyme that transcribes most of the genes in the human genome, operates as part of a large protein complex with multiple regulatory and accessory subunits.

Genetic Recombination

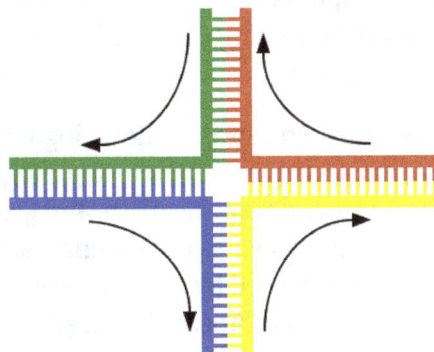

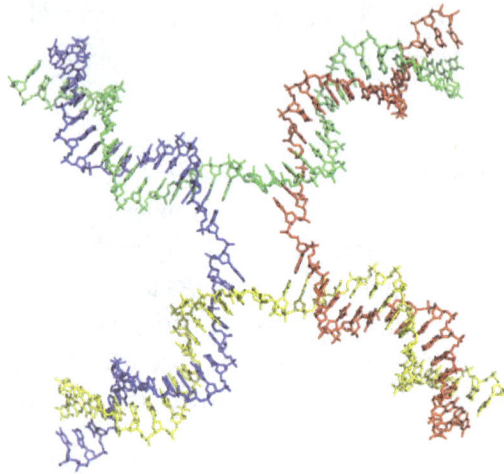

Structure of the Holliday junction intermediate in genetic recombination. The four separate DNA strands are coloured red, blue, green and yellow.

A DNA helix usually does not interact with other segments of DNA, and in human cells the different chromosomes even occupy separate areas in the nucleus called "chromosome territories". This physical separation of different chromosomes is important for the ability of DNA to function as a stable repository for information, as one of the few times chromosomes interact is in chromosomal crossover which occurs during sexual reproduction, when genetic recombination occurs. Chromosomal crossover is when two DNA helices break, swap a section and then rejoin.

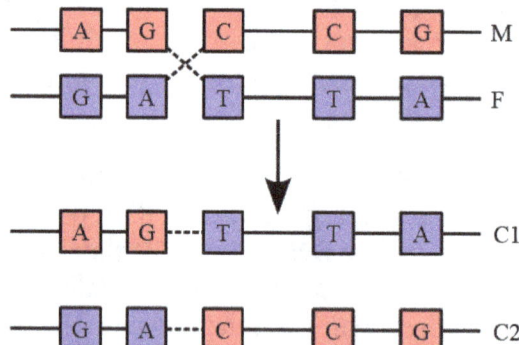

Recombination involves the breakage and rejoining of two chromosomes (M and F) to produce two re-arranged chromosomes (C1 and C2).

Recombination allows chromosomes to exchange genetic information and produces new combinations of genes, which increases the efficiency of natural selection and can be important in the rapid evolution of new proteins. Genetic recombination can also be involved in DNA repair, particularly in the cell's response to double-strand breaks.

The most common form of chromosomal crossover is homologous recombination, where the two chromosomes involved share very similar sequences. Non-homologous recombination can be damaging to cells, as it can produce chromosomal translocations and genetic abnormalities. The recombination reaction is catalyzed by enzymes known as recombinases, such as RAD51. The first step in recombination is a double-stranded break caused by either an endonuclease or damage to the DNA. A series of steps catalyzed in part by the recombinase then leads to joining of the two helices by at least one Holliday junction, in which a segment of a single strand in each helix is annealed to the complementary strand in the other helix. The Holliday junction is a tetrahedral junction structure that can be moved along the pair of chromosomes, swapping one strand for another. The recombination reaction is then halted by cleavage of the junction and re-ligation of the released DNA.

Evolution

DNA contains the genetic information that allows all modern living things to function, grow and reproduce. However, it is unclear how long in the 4-billion-year history of life DNA has performed this function, as it has been proposed that the earliest forms of life may have used RNA as their genetic material. RNA may have acted as the central part of early cell metabolism as it can both transmit genetic information and carry out catalysis as part of ribozymes. This ancient RNA world where nucleic acid would have been used for both catalysis and genetics may have influenced the evolution of the current genetic code based on four nucleotide bases. This would occur, since the number of different bases in such an organism is a trade-off between a small number of bases increasing replication accuracy and a large number of bases increasing the catalytic efficiency of ribozymes. However, there is no direct evidence of ancient genetic systems, as recovery of DNA from most fossils is impossible because DNA survives in the environment for less than one million years, and slowly degrades into short fragments in solution. Claims for older DNA have been made, most notably a report of the isolation of a viable bacterium from a salt crystal 250 million years old, but these claims are controversial.

Building blocks of DNA (adenine, guanine and related organic molecules) may have been formed extraterrestrially in outer space. Complex DNA and RNA organic compounds of life, including uracil, cytosine and thymine, have also been formed in the laboratory under conditions mimicking those found in outer space, using starting chemicals, such as pyrimidine, found in meteorites. Pyrimidine, like polycyclic aromatic hydrocarbons (PAHs), the most carbon-rich chemical found in the universe, may have been formed in red giants or in interstellar dust and gas clouds.

Uses in Technology

Genetic Engineering

Methods have been developed to purify DNA from organisms, such as phenol-chloroform extraction, and to manipulate it in the laboratory, such as restriction digests and the polymerase chain reaction. Modern biology and biochemistry make intensive use of these techniques in recombinant DNA technology. Recombinant DNA is a man-made DNA sequence that has been assembled from other DNA sequences. They can be transformed into organisms in the form of plasmids or in the appropriate format, by using a viral vector. The genetically modified organisms produced can be used to produce products such as recombinant proteins, used in medical research, or be grown in agriculture.

DNA Profiling

Forensic scientists can use DNA in blood, semen, skin, saliva or hair found at a crime scene to identify a matching DNA of an individual, such as a perpetrator. This process is formally termed DNA profiling, but may also be called "genetic fingerprinting". In DNA profiling, the lengths of variable sections of repetitive DNA, such as short tandem repeats and minisatellites, are compared between people. This method is usually an extremely reliable technique for identifying a matching DNA. However, identification can be complicated if the scene is contaminated with DNA from several people. DNA profiling was developed in 1984 by British geneticist Sir Alec Jeffreys, and first used in forensic science to convict Colin Pitchfork in the 1988 Enderby murders case.

The development of forensic science, and the ability to now obtain genetic matching on minute samples of blood, skin, saliva or hair has led to a re-examination of a number of cases. Evidence can now be uncovered that was not scientifically possible at the time of the original examination. Combined with the removal of the double jeopardy law in some places, this can allow cases to be reopened where previous trials have failed to produce sufficient evidence to convince a jury. People charged with serious crimes may be required to provide a sample of DNA for matching purposes. The most obvious defence to DNA matches obtained forensically is to claim that cross-contamination of evidence has taken place. This has resulted in meticulous strict handling procedures with new cases of serious crime. DNA profiling is also used to identify victims of mass casualty incidents. As well as positively identifying bodies or body parts in serious accidents, DNA profiling is being successfully used to identify individual victims in mass war graves – matching to family members.

DNA profiling is also used in DNA paternity testing in order to determine if someone is the biological parent or grandparent of a child with the probability of parentage is typically 99.99% when the alleged parent is biologically related to the child. Normal DNA sequencing methods happen after birth but there are new methods to test paternity while the mother is still pregnant.

DNA Enzymes or Catalytic DNA

Deoxyribozymes, also called DNAzymes or catalytic DNA are first discovered in 1994. They are mostly single stranded DNA sequences isolated from a large pool of random DNA sequences through a combinatorial approach called in vitro selection or SELEX. DNAzymes catalyze variety of chemical reactions including RNA/DNA cleavage, RNA/DNA ligation, amino acids phosphorylation/dephosphorylation, carbon-carbon bond formation, and etc. DNAzymes can enhance catalytic rate of chemical reactions up to 100,000,000,000-fold over the uncatalyzed reaction. The most extensively studied class of DNAzymes are RNA-cleaving DNAzymes which have been used in detection of different metal ions and designing therapeutic agents. Several metal-specific DNAzymes have been reported including the GR-5 DNAzyme (lead-specific), the CA1-3 DNAzymes (copper-specific), the 39E DNAzyme (uranyl-specific) and the NaA43 DNAzyme (sodium-specific). The NaA43 DNAzyme, which is reported to be more than 10,000-fold selective for sodium over other metal ions, was used to make a real-time sodium sensor in living cells.

Bioinformatics

Bioinformatics involves the development of techniques to store, data mine, search and manipulate biological data, including DNA nucleic acid sequence data. These have led to widely applied advances in computer science, especially string searching algorithms, machine learning and database theory. String searching or matching algorithms, which find an occurrence of a sequence of letters inside a larger sequence of letters, were developed to search for specific sequences of nucleotides. The DNA sequence may be aligned with other DNA sequences to identify homologous sequences and locate the specific mutations that make them distinct. These techniques, especially multiple sequence alignment, are used in studying phylogenetic relationships and protein function. Data sets representing entire genomes' worth of DNA sequences, such as those produced by the Human Genome Project, are difficult to use without the annotations that identify the locations of genes and regulatory elements on each chromosome. Regions of DNA sequence that have the characteristic patterns associated with protein- or RNA-coding genes can be identified by gene finding algorithms, which allow researchers to predict the presence of particular gene products and their possible functions in an organism even before they have been isolated experimentally. Entire genomes may also be compared, which can shed light on the evolutionary history of particular organism and permit the examination of complex evolutionary events.

DNA Nanotechnology

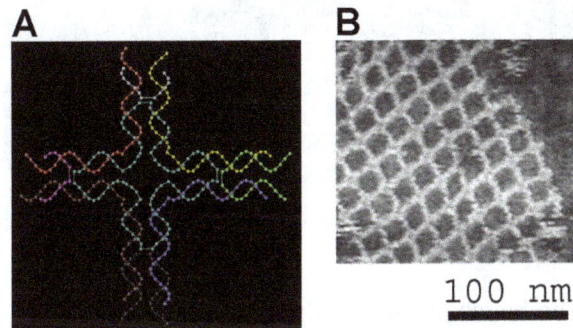

The DNA structure at left (schematic shown) will self-assemble into the structure visualized by atomic force microscopy at right. DNA nanotechnology is the field that seeks to design nanoscale structures using the molecular recognition properties of DNA molecules. Image from Strong, 2004.

DNA nanotechnology uses the unique molecular recognition properties of DNA and other nucleic acids to create self-assembling branched DNA complexes with useful properties. DNA is thus used as a structural material rather than as a carrier of biological information. This has led to the creation of two-dimensional periodic lattices (both tile-based and using the "DNA origami" method) as well as three-dimensional structures in the shapes of polyhedra. Nanomechanical devices and algorithmic self-assembly have also been demonstrated, and these DNA structures have been used to template the arrangement of other molecules such as gold nanoparticles and streptavidin proteins.

History and Anthropology

Because DNA collects mutations over time, which are then inherited, it contains historical information, and, by comparing DNA sequences, geneticists can infer the evolutionary history of organisms, their phylogeny. This field of phylogenetics is a powerful tool in evolutionary biology. If DNA sequences within a species are compared, population geneticists can learn the history of particular populations. This can be used in studies ranging from ecological genetics to anthropology; For example, DNA evidence is being used to try to identify the Ten Lost Tribes of Israel.

Information Storage

In a paper published in Nature in January 2013, scientists from the European Bioinformatics Institute and Agilent Technologies proposed a mechanism to use DNA's ability to code information as a means of digital data storage. The group was able to encode 739 kilobytes of data into DNA code, synthesize the actual DNA, then sequence the DNA and decode the information back to its original form, with a reported 100% accuracy. The encoded information consisted of text files and audio files. A prior experiment was published in August 2012. It was conducted by researchers at Harvard University, where the text of a 54,000-word book was encoded in DNA.

History of DNA Research

James Watson and Francis Crick (right), co-originators of the double-helix model, with Maclyn McCarty (left).

DNA was first isolated by the Swiss physician Friedrich Miescher who, in 1869, discovered a microscopic substance in the pus of discarded surgical bandages. As it resided in the nuclei of cells, he called it "nuclein". In 1878, Albrecht Kossel isolated the non-protein component of "nuclein", nucleic acid, and later isolated its five primary nucleobases. In 1919, Phoebus Levene identified the base, sugar and phosphate nucleotide unit. Levene suggested that DNA consisted of a string of nucleotide units linked together through the phosphate groups. Levene thought the chain was short and the bases repeated in a fixed order. In 1937, William Astbury produced the first X-ray diffraction patterns that showed that DNA had a regular structure.

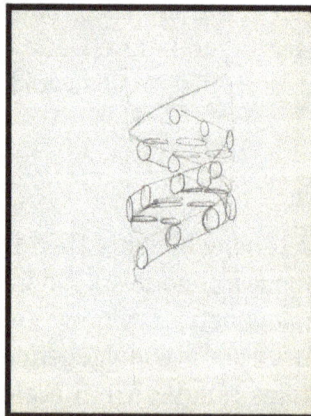

Pencil sketch of the DNA double helix by Francis Crick in 1953

In 1927, Nikolai Koltsov proposed that inherited traits would be inherited via a "giant hereditary molecule" made up of "two mirror strands that would replicate in a semi-conservative fashion using each strand as a template". In 1928, Frederick Griffith in his experiment discovered that traits of the "smooth" form of Pneumococcus could be transferred to the "rough" form of the same bacteria by mixing killed "smooth" bacteria with the live "rough" form. This system provided the first clear suggestion that DNA carries genetic information—the Avery–MacLeod–McCarty experiment—when Oswald Avery, along with coworkers Colin MacLeod and Maclyn McCarty, identified DNA as the transforming principle in 1943. DNA's role in heredity was confirmed in 1952, when

Alfred Hershey and Martha Chase in the Hershey–Chase experiment showed that DNA is the genetic material of the T2 phage.

In 1953, James Watson and Francis Crick suggested what is now accepted as the first correct double-helix model of DNA structure in the journal Nature. Their double-helix, molecular model of DNA was then based on a single X-ray diffraction image (labeled as "Photo 51") taken by Rosalind Franklin and Raymond Gosling in May 1952, as well as the information that the DNA bases are paired.

Experimental evidence supporting the Watson and Crick model was published in a series of five articles in the same issue of Nature. Of these, Franklin and Gosling's paper was the first publication of their own X-ray diffraction data and original analysis method that partially supported the Watson and Crick model; this issue also contained an article on DNA structure by Maurice Wilkins and two of his colleagues, whose analysis and in vivo B-DNA X-ray patterns also supported the presence in vivo of the double-helical DNA configurations as proposed by Crick and Watson for their double-helix molecular model of DNA in the previous two pages of Nature. In 1962, after Franklin's death, Watson, Crick, and Wilkins jointly received the Nobel Prize in Physiology or Medicine. Nobel Prizes are awarded only to living recipients. A debate continues about who should receive credit for the discovery.

In an influential presentation in 1957, Crick laid out the central dogma of molecular biology, which foretold the relationship between DNA, RNA, and proteins, and articulated the "adaptor hypothesis". Final confirmation of the replication mechanism that was implied by the double-helical structure followed in 1958 through the Meselson–Stahl experiment. Further work by Crick and coworkers showed that the genetic code was based on non-overlapping triplets of bases, called codons, allowing Har Gobind Khorana, Robert W. Holley and Marshall Warren Nirenberg to decipher the genetic code. These findings represent the birth of molecular biology.

RNA

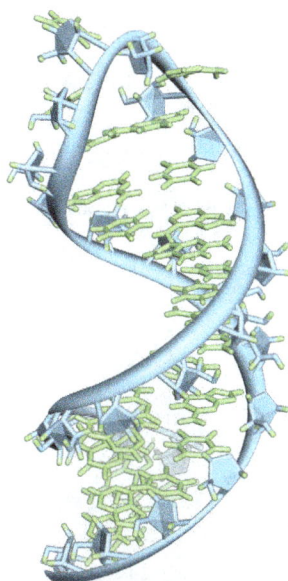

A hairpin loop from a pre-mRNA. Highlighted are the nucleobases (green) and the ribose-phosphate backbone (blue). Note that this is a single strand of RNA that folds back upon itself.

Ribonucleic acid (RNA) is a polymeric molecule implicated in various biological roles in coding, decoding, regulation, and expression of genes. RNA and DNA are nucleic acids, and, along with proteins and carbohydrates, constitute the three major macromolecules essential for all known forms of life. Like DNA, RNA is assembled as a chain of nucleotides, but unlike DNA it is more often found in nature as a single-strand folded onto itself, rather than a paired double-strand. Cellular organisms use messenger RNA (mRNA) to convey genetic information (using the letters G, U, A, and C to denote the nitrogenous bases guanine, uracil, adenine, and cytosine) that directs synthesis of specific proteins. Many viruses encode their genetic information using an RNA genome.

Some RNA molecules play an active role within cells by catalyzing biological reactions, controlling gene expression, or sensing and communicating responses to cellular signals. One of these active processes is protein synthesis, a universal function wherein mRNA molecules direct the assembly of proteins on ribosomes. This process uses transfer RNA (tRNA) molecules to deliver amino acids to the ribosome, where ribosomal RNA (rRNA) then links amino acids together to form proteins.

Comparison with DNA

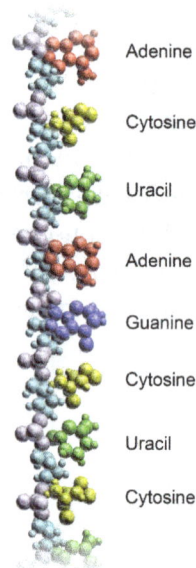

Adenine

Cytosine

Uracil

Adenine

Guanine

Cytosine

Uracil

Cytosine

RNA Molecule

Bases in an RNA Molecule.

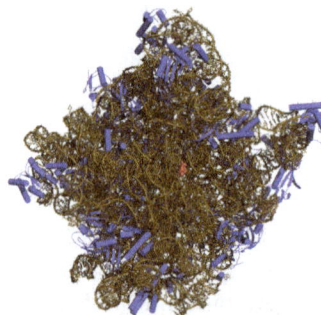

Three-dimensional representation of the 50S ribosomal subunit. Ribosomal RNA is in ochre, proteins in blue. The active site is a small segment of rRNA, indicated in red.

The chemical structure of RNA is very similar to that of DNA, but differs in three main ways:

- Unlike double-stranded DNA, RNA is a single-stranded molecule in many of its biological roles and has a much shorter chain of nucleotides. However, RNA can, by complementary base pairing, form intrastrand (i.e., single-strand) double helixes, as in tRNA.

- While DNA contains deoxyribose, RNA contains ribose (in deoxyribose there is no hydroxyl group attached to the pentose ring in the 2' position). These hydroxyl groups make RNA less stable than DNA because it is more prone to hydrolysis.

- The complementary base to adenine in DNA is thymine, whereas in RNA, it is uracil, which is an unmethylated form of thymine.

Like DNA, most biologically active RNAs, including mRNA, tRNA, rRNA, snRNAs, and other non-coding RNAs, contain self-complementary sequences that allow parts of the RNA to fold and pair with itself to form double helices. Analysis of these RNAs has revealed that they are highly structured. Unlike DNA, their structures do not consist of long double helices, but rather collections of short helices packed together into structures akin to proteins. In this fashion, RNAs can achieve chemical catalysis (like enzymes). For instance, determination of the structure of the ribosome—an enzyme that catalyzes peptide bond formation—revealed that its active site is composed entirely of RNA.

Structure

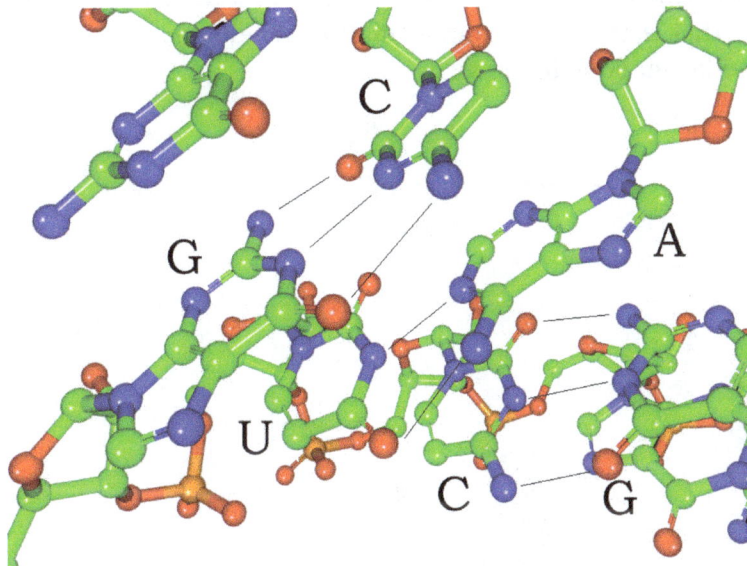

Watson-Crick base pairs in a siRNA (hydrogen atoms are not shown)

Each nucleotide in RNA contains a ribose sugar, with carbons numbered 1' through 5'. A base is attached to the 1' position, in general, adenine (A), cytosine (C), guanine (G), or uracil (U). Adenine and guanine are purines, cytosine and uracil are pyrimidines. A phosphate group is attached to the 3' position of one ribose and the 5' position of the next. The phosphate groups have a negative charge each, making RNA a charged molecule (polyanion). The bases form hydrogen bonds between cytosine and guanine, between adenine and uracil and between guanine and uracil. However, other interactions are possible, such as a group of adenine bases binding to each other in a bulge, or the GNRA tetraloop that has a guanine–adenine base-pair.

Chemical structure of RNA

An important structural feature of RNA that distinguishes it from DNA is the presence of a hydroxyl group at the 2' position of the ribose sugar. The presence of this functional group causes the helix to mostly adopt the A-form geometry, although in single strand dinucleotide contexts, RNA can rarely also adopt the B-form most commonly observed in DNA. The A-form geometry results in a very deep and narrow major groove and a shallow and wide minor groove. A second consequence of the presence of the 2'-hydroxyl group is that in conformationally flexible regions of an RNA molecule (that is, not involved in formation of a double helix), it can chemically attack the adjacent phosphodiester bond to cleave the backbone.

Secondary structure of a telomerase RNA.

RNA is transcribed with only four bases (adenine, cytosine, guanine and uracil), but these bases and attached sugars can be modified in numerous ways as the RNAs mature. Pseudouridine (Ψ), in which the linkage between uracil and ribose is changed from a C–N bond to a C–C bond, and ribothymidine (T) are found in various places (the most notable ones being in the TΨC loop of tRNA). Another notable modified base is hypoxanthine, a deaminated adenine base whose nucleoside is called inosine (I). Inosine plays a key role in the wobble hypothesis of the genetic code.

There are more than 100 other naturally occurring modified nucleosides, The greatest structural diversity of modifications can be found in tRNA, while pseudouridine and nucleosides with 2'-O-methylribose often present in rRNA are the most common. The specific roles of many of these modifications in RNA are not fully understood. However, it is notable that, in ribosomal RNA, many of the post-transcriptional modifications occur in highly functional regions, such as the peptidyl transferase center and the subunit interface, implying that they are important for normal function.

The functional form of single-stranded RNA molecules, just like proteins, frequently requires a specific tertiary structure. The scaffold for this structure is provided by secondary structural elements that are hydrogen bonds within the molecule. This leads to several recognizable "domains" of secondary structure like hairpin loops, bulges, and internal loops. Since RNA is charged, metal ions such as Mg^+ are needed to stabilise many secondary and tertiary structures.

The naturally occurring enantiomer of RNA is D-RNA composed of D-ribonucleotides. All chirality centers are located in the D-ribose. By the use of L-ribose or rather L-ribonucleotides, L-RNA can be synthesized. L-RNA is much more stable against degradation by RNase.

Like other structured biopolymers such as proteins, one can define topology of a folded RNA molecule. This is often done based on arrangement of intra-chain contacts within a folded RNA, termed as circuit topology.

Synthesis

Synthesis of RNA is usually catalyzed by an enzyme—RNA polymerase—using DNA as a template, a process known as transcription. Initiation of transcription begins with the binding of the enzyme to a promoter sequence in the DNA (usually found "upstream" of a gene). The DNA double helix is unwound by the helicase activity of the enzyme. The enzyme then progresses along the template strand in the 3' to 5' direction, synthesizing a complementary RNA molecule with elongation occurring in the 5' to 3' direction. The DNA sequence also dictates where termination of RNA synthesis will occur.

Primary transcript RNAs are often modified by enzymes after transcription. For example, a poly(A) tail and a 5' cap are added to eukaryotic pre-mRNA and introns are removed by the spliceosome.

There are also a number of RNA-dependent RNA polymerases that use RNA as their template for synthesis of a new strand of RNA. For instance, a number of RNA viruses (such as poliovirus) use this type of enzyme to replicate their genetic material. Also, RNA-dependent RNA polymerase is part of the RNA interference pathway in many organisms.

Types of RNA

Overview

Messenger RNA (mRNA) is the RNA that carries information from DNA to the ribosome, the sites of protein synthesis (translation) in the cell. The coding sequence of the mRNA determines the amino acid sequence in the protein that is produced. However, many RNAs do not code for protein (about 97% of the transcriptional output is non-protein-coding in eukaryotes).

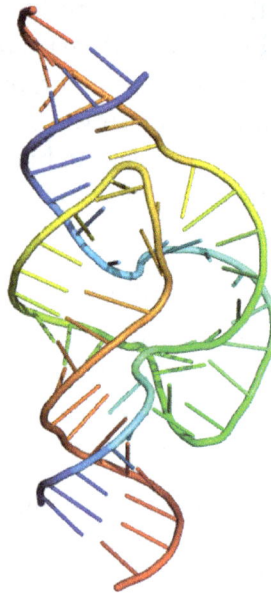

Structure of a hammerhead ribozyme, a ribozyme that cuts RNA

These so-called non-coding RNAs ("ncRNA") can be encoded by their own genes (RNA genes), but can also derive from mRNA introns. The most prominent examples of non-coding RNAs are transfer RNA (tRNA) and ribosomal RNA (rRNA), both of which are involved in the process of translation. There are also non-coding RNAs involved in gene regulation, RNA processing and other roles. Certain RNAs are able to catalyse chemical reactions such as cutting and ligating other RNA molecules, and the catalysis of peptide bond formation in the ribosome; these are known as ribozymes.

In Length

According to the length of RNA chain, RNA includes small RNA and long RNA. Usually, small RNAs are <200 nt in length, and long RNA are >200 nt. Long RNAs, also called large RNAs, mainly include long non-coding RNA (lncRNA) and mRNA. Small RNAs mainly include 5.8S ribosomal RNA (rRNA), 5S rRNA, transfer RNA (tRNA), microRNA (miRNA), small interfering RNA (siRNA), small nucleolar RNA (snoRNAs), Piwi-interacting RNA (piRNA), tRNA-derived small RNA (tsRNA) and small rDNA-derived RNA (srRNA).

In Translation

Messenger RNA (mRNA) carries information about a protein sequence to the ribosomes, the protein synthesis factories in the cell. It is coded so that every three nucleotides (a codon) correspond to one amino acid. In eukaryotic cells, once precursor mRNA (pre-mRNA) has been transcribed from DNA, it is processed to mature mRNA. This removes its introns—non-coding sections of the pre-mRNA. The mRNA is then exported from the nucleus to the cytoplasm, where it is bound to ribosomes and translated into its corresponding protein form with the help of tRNA. In prokaryotic cells, which do not have nucleus and cytoplasm compartments, mRNA can bind to ribosomes while it is being transcribed from DNA. After a certain amount of time the message degrades into its component nucleotides with the assistance of ribonucleases.

Transfer RNA (tRNA) is a small RNA chain of about 80 nucleotides that transfers a specific amino acid to a growing polypeptide chain at the ribosomal site of protein synthesis during translation. It has sites for amino acid attachment and an anticodon region for codon recognition that binds to a specific sequence on the messenger RNA chain through hydrogen bonding.

Ribosomal RNA (rRNA) is the catalytic component of the ribosomes. Eukaryotic ribosomes contain four different rRNA molecules: 18S, 5.8S, 28S and 5S rRNA. Three of the rRNA molecules are synthesized in the nucleolus, and one is synthesized elsewhere. In the cytoplasm, ribosomal RNA and protein combine to form a nucleoprotein called a ribosome. The ribosome binds mRNA and carries out protein synthesis. Several ribosomes may be attached to a single mRNA at any time. Nearly all the RNA found in a typical eukaryotic cell is rRNA.

Transfer-messenger RNA (tmRNA) is found in many bacteria and plastids. It tags proteins encoded by mRNAs that lack stop codons for degradation and prevents the ribosome from stalling.

Regulatory RNAs

Several types of RNA can downregulate gene expression by being complementary to a part of an mRNA or a gene's DNA. MicroRNAs (miRNA; 21-22 nt) are found in eukaryotes and act through RNA interference (RNAi), where an effector complex of miRNA and enzymes can cleave complementary mRNA, block the mRNA from being translated, or accelerate its degradation.

While small interfering RNAs (siRNA; 20-25 nt) are often produced by breakdown of viral RNA, there are also endogenous sources of siRNAs. siRNAs act through RNA interference in a fashion similar to miRNAs. Some miRNAs and siRNAs can cause genes they target to be methylated, thereby decreasing or increasing transcription of those genes. Animals have Piwi-interacting RNAs (piRNA; 29-30 nt) that are active in germline cells and are thought to be a defense against transposons and play a role in gametogenesis.

Many prokaryotes have CRISPR RNAs, a regulatory system similar to RNA interference. Antisense RNAs are widespread; most downregulate a gene, but a few are activators of transcription. One way antisense RNA can act is by binding to an mRNA, forming double-stranded RNA that is enzymatically degraded. There are many long noncoding RNAs that regulate genes in eukaryotes, one such RNA is Xist, which coats one X chromosome in female mammals and inactivates it.

An mRNA may contain regulatory elements itself, such as riboswitches, in the 5' untranslated region or 3' untranslated region; these cis-regulatory elements regulate the activity of that mRNA. The untranslated regions can also contain elements that regulate other genes.

In RNA Processing

Many RNAs are involved in modifying other RNAs. Introns are spliced out of pre-mRNA by spliceosomes, which contain several small nuclear RNAs (snRNA), or the introns can be ribozymes that are spliced by themselves. RNA can also be altered by having its nucleotides modified to nucleotides other than A, C, G and U. In eukaryotes, modifications of RNA nucleotides are in general directed by small nucleolar RNAs (snoRNA; 60-300 nt), found in the nucleolus and cajal bodies. snoRNAs associate with enzymes and guide them to a spot on an RNA by basepairing to that

RNA. These enzymes then perform the nucleotide modification. rRNAs and tRNAs are extensively modified, but snRNAs and mRNAs can also be the target of base modification. RNA can also be methylated.

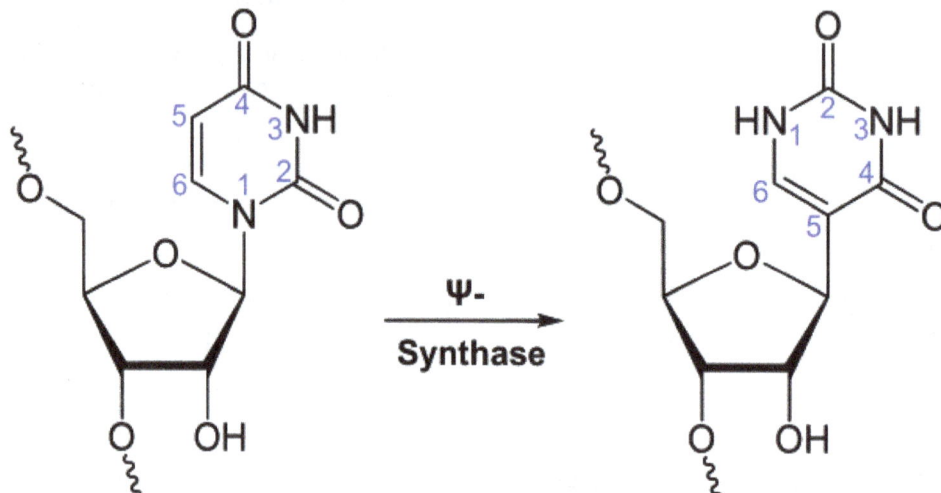

Uridine to pseudouridine is a common RNA modification.

RNA Genomes

Like DNA, RNA can carry genetic information. RNA viruses have genomes composed of RNA that encodes a number of proteins. The viral genome is replicated by some of those proteins, while other proteins protect the genome as the virus particle moves to a new host cell. Viroids are another group of pathogens, but they consist only of RNA, do not encode any protein and are replicated by a host plant cell's polymerase.

In reverse Transcription

Reverse transcribing viruses replicate their genomes by reverse transcribing DNA copies from their RNA; these DNA copies are then transcribed to new RNA. Retrotransposons also spread by copying DNA and RNA from one another, and telomerase contains an RNA that is used as template for building the ends of eukaryotic chromosomes.

Double-Stranded RNA

Double-stranded RNA (dsRNA) is RNA with two complementary strands, similar to the DNA found in all cells. dsRNA forms the genetic material of some viruses (double-stranded RNA viruses). Double-stranded RNA such as viral RNA or siRNA can trigger RNA interference in eukaryotes, as well as interferon response in vertebrates.

Circular RNA

Recently, it was shown that there is a single stranded covalently closed, i.e. circular form of RNA expressed throughout the animal and plant kingdom. circRNAs are thought to arise via a "back-splice" reaction where the spliceosome joins a downstream donor to an upstream acceptor splice site. So far the function of circRNAs is largely unknown, although for few examples a microRNA sponging activity has been demonstrated.

Key Discoveries in RNA Biology

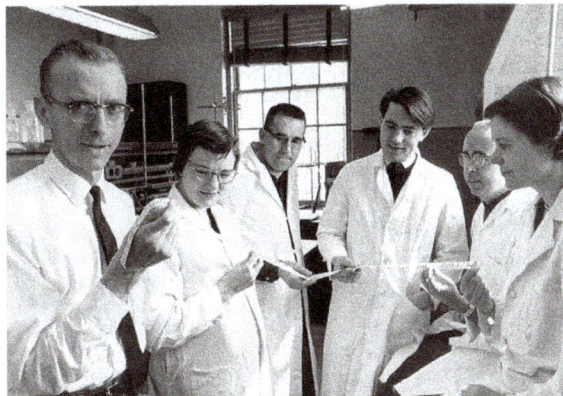

Robert W. Holley, left, poses with his research team.

Research on RNA has led to many important biological discoveries and numerous Nobel Prizes. Nucleic acids were discovered in 1868 by Friedrich Miescher, who called the material 'nuclein' since it was found in the nucleus. It was later discovered that prokaryotic cells, which do not have a nucleus, also contain nucleic acids. The role of RNA in protein synthesis was suspected already in 1939. Severo Ochoa won the 1959 Nobel Prize in Medicine (shared with Arthur Kornberg) after he discovered an enzyme that can synthesize RNA in the laboratory. However, the enzyme discovered by Ochoa (polynucleotide phosphorylase) was later shown to be responsible for RNA degradation, not RNA synthesis. In 1956 Alex Rich and David Davies hybridized two separate strands of RNA to form the first crystal of RNA whose structure could be determined by X-ray crystallography.

The sequence of the 77 nucleotides of a yeast tRNA was found by Robert W. Holley in 1965, winning Holley the 1968 Nobel Prize in Medicine (shared with Har Gobind Khorana and Marshall Nirenberg). In 1967, Carl Woese hypothesized that RNA might be catalytic and suggested that the earliest forms of life (self-replicating molecules) could have relied on RNA both to carry genetic information and to catalyze biochemical reactions—an RNA world.

During the early 1970s, retroviruses and reverse transcriptase were discovered, showing for the first time that enzymes could copy RNA into DNA (the opposite of the usual route for transmission of genetic information). For this work, David Baltimore, Renato Dulbecco and Howard Temin were awarded a Nobel Prize in 1975. In 1976, Walter Fiers and his team determined the first complete nucleotide sequence of an RNA virus genome, that of bacteriophage MS2.

In 1977, introns and RNA splicing were discovered in both mammalian viruses and in cellular genes, resulting in a 1993 Nobel to Philip Sharp and Richard Roberts. Catalytic RNA molecules (ribozymes) were discovered in the early 1980s, leading to a 1989 Nobel award to Thomas Cech and Sidney Altman. In 1990, it was found in Petunia that introduced genes can silence similar genes of the plant's own, now known to be a result of RNA interference.

At about the same time, 22 nt long RNAs, now called microRNAs, were found to have a role in the development of C. elegans. Studies on RNA interference gleaned a Nobel Prize for Andrew Fire and Craig Mello in 2006, and another Nobel was awarded for studies on the transcription of RNA to Roger Kornberg in the same year. The discovery of gene regulatory RNAs has led to attempts to develop drugs made of RNA, such as siRNA, to silence genes.

Evolution

In March 2015, complex DNA and RNA organic compounds of life, including uracil, cytosine and thymine, were reportedly formed in the laboratory under outer space conditions, using starting chemicals, such as pyrimidine, found in meteorites. Pyrimidine, like polycyclic aromatic hydrocarbons (PAHs), the most carbon-rich chemical found in the Universe, may have been formed in red giants or in interstellar dust and gas clouds, according to the scientists.

Protein

Proteins are large biomolecules, or macromolecules, consisting of one or more long chains of amino acid residues. Proteins perform a vast array of functions within organisms, including catalysing metabolic reactions, DNA replication, responding to stimuli, and transporting molecules from one location to another. Proteins differ from one another primarily in their sequence of amino acids, which is dictated by the nucleotide sequence of their genes, and which usually results in protein folding into a specific three-dimensional structure that determines its activity.

A linear chain of amino acid residues is called a polypeptide. A protein contains at least one long polypeptide. Short polypeptides, containing less than 20–30 residues, are rarely considered to be proteins and are commonly called peptides, or sometimes oligopeptides. The individual amino acid residues are bonded together by peptide bonds and adjacent amino acid residues. The sequence of amino acid residues in a protein is defined by the sequence of a gene, which is encoded in the genetic code. In general, the genetic code specifies 20 standard amino acids; however, in certain organisms the genetic code can include selenocysteine and—in certain archaea—pyrrolysine. Shortly after or even during synthesis, the residues in a protein are often chemically modified by post-translational modification, which alters the physical and chemical properties, folding, stability, activity, and ultimately, the function of the proteins. Sometimes proteins have non-peptide groups attached, which can be called prosthetic groups or cofactors. Proteins can also work together to achieve a particular function, and they often associate to form stable protein complexes.

Once formed, proteins only exist for a certain period of time and are then degraded and recycled by the cell's machinery through the process of protein turnover. A protein's lifespan is measured in terms of its half-life and covers a wide range. They can exist for minutes or years with an average lifespan of 1–2 days in mammalian cells. Abnormal and or misfolded proteins are degraded more rapidly either due to being targeted for destruction or due to being unstable.

Like other biological macromolecules such as polysaccharides and nucleic acids, proteins are essential parts of organisms and participate in virtually every process within cells. Many proteins are enzymes that catalyse biochemical reactions and are vital to metabolism. Proteins also have structural or mechanical functions, such as actin and myosin in muscle and the proteins in the cytoskeleton, which form a system of scaffolding that maintains cell shape. Other proteins are important in cell signaling, immune responses, cell adhesion, and the cell cycle. In animals, proteins are needed in the diet to provide the essential amino acids that cannot be synthesized. Digestion breaks the proteins down for use in the metabolism.

Proteins may be purified from other cellular components using a variety of techniques such as ultracentrifugation, precipitation, electrophoresis, and chromatography; the advent of genetic engineering has made possible a number of methods to facilitate purification. Methods commonly used to study protein structure and function include immunohistochemistry, site-directed mutagenesis, X-ray crystallography, nuclear magnetic resonance and mass spectrometry.

Biochemistry

Chemical structure of the peptide bond (bottom) and the three-dimensional structure of a peptide bond between an alanine and an adjacent amino acid (top/inset)

Resonance structures of the peptide bond that links individual amino acids to form a protein polymer

Most proteins consist of linear polymers built from series of up to 20 different L-α-amino acids. All proteinogenic amino acids possess common structural features, including an α-carbon to which an amino group, a carboxyl group, and a variable side chain are bonded. Only proline differs from this basic structure as it contains an unusual ring to the N-end amine group, which forces the CO−NH amide moiety into a fixed conformation. The side chains of the standard amino acids, detailed in the list of standard amino acids, have a great variety of chemical structures and properties; it is the combined effect of all of the amino acid side chains in a protein that ultimately determines its three-dimensional structure and its chemical reactivity. The amino acids in a polypeptide chain are linked by peptide bonds. Once linked in the protein chain, an individual amino acid is called a residue, and the linked series of carbon, nitrogen, and oxygen atoms are known as the main chain or protein backbone.

The peptide bond has two resonance forms that contribute some double-bond character and inhibit rotation around its axis, so that the alpha carbons are roughly coplanar. The other two dihedral angles in the peptide bond determine the local shape assumed by the protein backbone. The end of the protein with a free carboxyl group is known as the C-terminus or carboxy terminus, whereas the end with a free amino group is known as the N-terminus or amino terminus. The words protein, polypeptide, and peptide are a little ambiguous and can overlap in meaning. Protein is generally used to refer to the complete biological molecule in a stable conformation, whereas peptide is generally reserved for a short amino acid oligomers often lacking a stable three-dimensional structure. However, the boundary between the two is not well defined and usually lies near 20–30 residues. Polypeptide can refer to any single linear chain of amino acids, usually regardless of length, but often implies an absence of a defined conformation.

Abundance in Cells

It has been estimated that average-sized bacteria contain about 2 million proteins per cell (e.g. E. coli and Staphylococcus aureus). Smaller bacteria, such as Mycoplasma or spirochetes contain fewer molecules, namely on the order of 50,000 to 1 million. By contrast, eukaryotic cells are larger and thus contain much more protein. For instance, yeast cells were estimated to contain about 50 million proteins and human cells on the order of 1 to 3 billion. Bacterial genomes encode about 10 times fewer proteins than humans (e.g. small bacteria ~1,000, E. coli: ~4,000, yeast: ~6,000, human: ~20,000).

The concentration of individual proteins ranges from a few molecules per cell to hundreds of thousands, and about a third of all proteins is not produced in most cells or only induced under certain circumstances. For instance, of the 20,000 or so proteins encoded by the human genome only 6,000 are detected in lymphoblastoid cells.

Synthesis

Biosynthesis

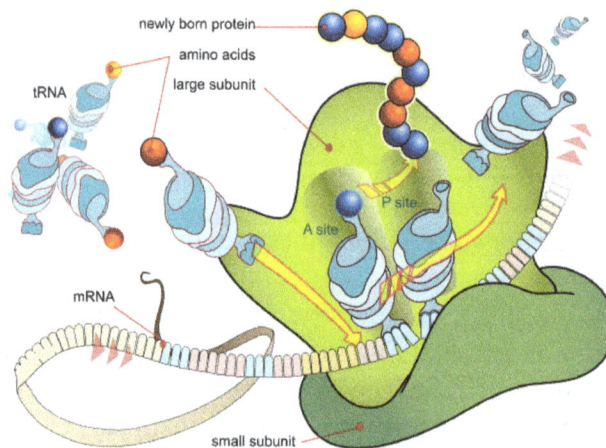

A ribosome produces a protein using mRNA as template

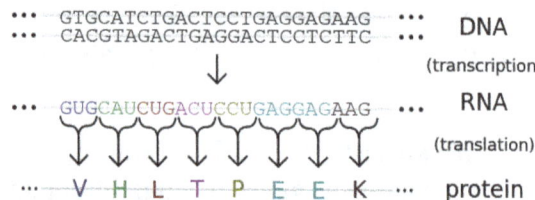

The DNA sequence of a gene encodes the amino acid sequence of a protein

Proteins are assembled from amino acids using information encoded in genes. Each protein has its own unique amino acid sequence that is specified by the nucleotide sequence of the gene encoding this protein. The genetic code is a set of three-nucleotide sets called codons and each three-nucleotide combination designates an amino acid, for example AUG (adenine-uracil-guanine) is the code for methionine. Because DNA contains four nucleotides, the total number of possible codons is 64; hence, there is some redundancy in the genetic code, with some amino acids specified by more than one codon. Genes encoded in DNA are first transcribed into pre-messenger RNA (mRNA) by proteins such as RNA polymerase. Most organisms then process the pre-mRNA (also known as a primary transcript) using various forms of Post-transcriptional modification to form the mature mRNA, which is then used as a template for protein synthesis by the ribosome. In prokaryotes the mRNA may either be used as soon as it is produced, or be bound by a ribosome after having moved away from the nucleoid. In contrast, eukaryotes make mRNA in the cell nucleus and then translocate it across the nuclear membrane into the cytoplasm, where protein synthesis then takes place. The rate of protein synthesis is higher in prokaryotes than eukaryotes and can reach up to 20 amino acids per second.

The process of synthesizing a protein from an mRNA template is known as translation. The mRNA is loaded onto the ribosome and is read three nucleotides at a time by matching each codon to its base pairing anticodon located on a transfer RNA molecule, which carries the amino acid corresponding to the codon it recognizes. The enzyme aminoacyl tRNA synthetase "charges" the tRNA molecules with the correct amino acids. The growing polypeptide is often termed the nascent chain. Proteins are always biosynthesized from N-terminus to C-terminus.

The size of a synthesized protein can be measured by the number of amino acids it contains and by its total molecular mass, which is normally reported in units of daltons (synonymous with atomic mass units), or the derivative unit kilodalton (kDa). Yeast proteins are on average 466 amino acids long and 53 kDa in mass. The largest known proteins are the titins, a component of the muscle sarcomere, with a molecular mass of almost 3,000 kDa and a total length of almost 27,000 amino acids.

Chemical Synthesis

Short proteins can also be synthesized chemically by a family of methods known as peptide synthesis, which rely on organic synthesis techniques such as chemical ligation to produce peptides in high yield. Chemical synthesis allows for the introduction of non-natural amino acids into polypeptide chains, such as attachment of fluorescent probes to amino acid side chains. These methods are useful in laboratory biochemistry and cell biology, though generally not for commercial applications. Chemical synthesis is inefficient for polypeptides longer than about 300 amino acids, and the synthesized proteins may not readily assume their native tertiary structure. Most chemical synthesis methods proceed from C-terminus to N-terminus, opposite the biological reaction.

Structure

Most proteins fold into unique 3-dimensional structures. The shape into which a protein naturally folds is known as its native conformation. Although many proteins can fold unassisted, simply through the chemical properties of their amino acids, others require the aid of molecular chaperones to fold into their native states. Biochemists often refer to four distinct aspects of a protein's structure:

- Primary structure: the amino acid sequence. A protein is a polyamide.

- Secondary structure: regularly repeating local structures stabilized by hydrogen bonds.

The most common examples are the α-helix, β-sheet and turns. Because secondary structures are local, many regions of different secondary structure can be present in the same protein molecule.

- Tertiary structure: the overall shape of a single protein molecule; the spatial relationship of the secondary structures to one another. Tertiary structure is generally stabilized by nonlocal interactions, most commonly the formation of a hydrophobic core, but also through salt bridges, hydrogen bonds, disulfide bonds, and even posttranslational modifications. The term "tertiary structure" is often used as synonymous with the term fold. The tertiary structure is what controls the basic function of the protein.

- Quaternary structure: the structure formed by several protein molecules (polypeptide chains), usually called protein subunits in this context, which function as a single protein complex.

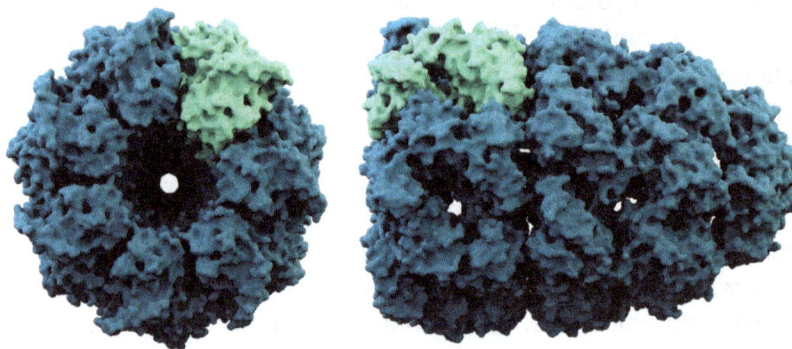

The crystal structure of the chaperonin, a huge protein complex. A single protein subunit is highlighted. Chaperonins assist protein folding.

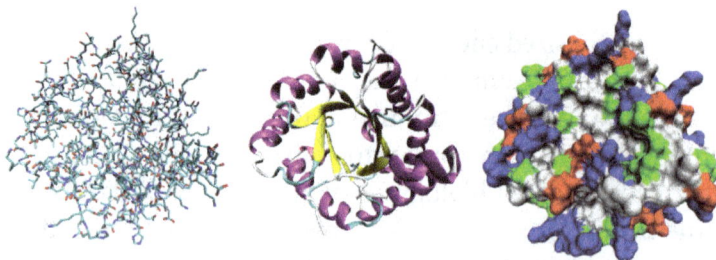

Three possible representations of the three-dimensional structure of the protein triose phosphate isomerase. Left: All-atom representation colored by atom type. Middle: Simplified representation illustrating the backbone conformation, colored by secondary structure. Right: Solvent-accessible surface representation colored by residue type (acidic residues red, basic residues blue, polar residues green, nonpolar residues white).

Proteins are not entirely rigid molecules. In addition to these levels of structure, proteins may shift between several related structures while they perform their functions. In the context of these functional rearrangements, these tertiary or quaternary structures are usually referred to as "conformations", and transitions between them are called conformational changes. Such changes are often induced by the binding of a substrate molecule to an enzyme's active site, or the physical region of the protein that participates in chemical catalysis. In solution proteins also undergo variation in structure through thermal vibration and the collision with other molecules.

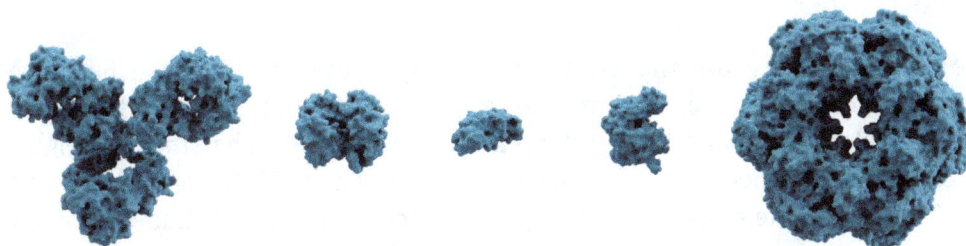

Molecular surface of several proteins showing their comparative sizes. From left to right are: immunoglobulin G (IgG, an antibody), hemoglobin, insulin (a hormone), adenylate kinase (an enzyme), and glutamine synthetase (an enzyme).

Proteins can be informally divided into three main classes, which correlate with typical tertiary structures: globular proteins, fibrous proteins, and membrane proteins. Almost all globular proteins are soluble and many are enzymes. Fibrous proteins are often structural, such as collagen, the major component of connective tissue, or keratin, the protein component of hair and nails. Membrane proteins often serve as receptors or provide channels for polar or charged molecules to pass through the cell membrane.

A special case of intramolecular hydrogen bonds within proteins, poorly shielded from water attack and hence promoting their own dehydration, are called dehydrons.

Structure Determination

Discovering the tertiary structure of a protein, or the quaternary structure of its complexes, can provide important clues about how the protein performs its function. Common experimental methods of structure determination include X-ray crystallography and NMR spectroscopy, both of which can produce information at atomic resolution. However, NMR experiments are able to provide information from which a subset of distances between pairs of atoms can be estimated, and the final possible conformations for a protein are determined by solving a distance geometry problem. Dual polarisation interferometry is a quantitative analytical method for measuring the overall protein conformation and conformational changes due to interactions or other stimulus. Circular dichroism is another laboratory technique for determining internal β-sheet / α-helical composition of proteins. Cryoelectron microscopy is used to produce lower-resolution structural information about very large protein complexes, including assembled viruses; a variant known as electron crystallography can also produce high-resolution information in some cases, especially for two-dimensional crystals of membrane proteins. Solved structures are usually deposited in the Protein Data Bank (PDB), a freely available resource from which structural data about thousands of proteins can be obtained in the form of Cartesian coordinates for each atom in the protein.

Many more gene sequences are known than protein structures. Further, the set of solved structures is biased toward proteins that can be easily subjected to the conditions required in X-ray crystallography, one of the major structure determination methods. In particular, globular proteins are comparatively easy to crystallize in preparation for X-ray crystallography. Membrane proteins, by contrast, are difficult to crystallize and are underrepresented in the PDB. Structural genomics initiatives have attempted to remedy these deficiencies by systematically solving representative structures of major fold classes. Protein structure prediction methods attempt to provide a means of generating a plausible structure for proteins whose structures have not been experimentally determined.

Cellular Functions

Proteins are the chief actors within the cell, said to be carrying out the duties specified by the information encoded in genes. With the exception of certain types of RNA, most other biological molecules are relatively inert elements upon which proteins act. Proteins make up half the dry weight of an Escherichia coli cell, whereas other macromolecules such as DNA and RNA make up only 3% and 20%, respectively. The set of proteins expressed in a particular cell or cell type is known as its proteome.

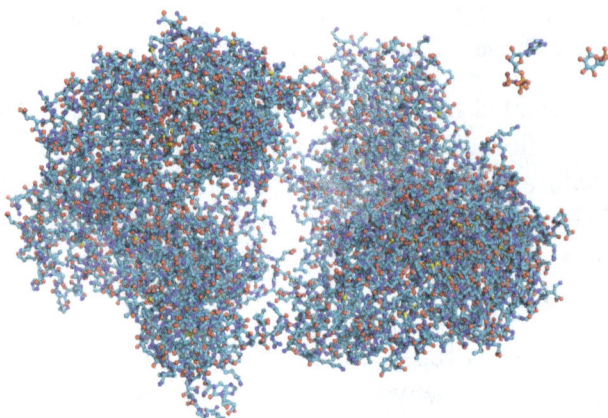

The enzyme hexokinase is shown as a conventional ball-and-stick molecular model. To scale in the top right-hand corner are two of its substrates, ATP and glucose.

The chief characteristic of proteins that also allows their diverse set of functions is their ability to bind other molecules specifically and tightly. The region of the protein responsible for binding another molecule is known as the binding site and is often a depression or "pocket" on the molecular surface. This binding ability is mediated by the tertiary structure of the protein, which defines the binding site pocket, and by the chemical properties of the surrounding amino acids' side chains. Protein binding can be extraordinarily tight and specific; for example, the ribonuclease inhibitor protein binds to human angiogenin with a sub-femtomolar dissociation constant ($<10^-$ M) but does not bind at all to its amphibian homolog onconase (>1 M). Extremely minor chemical changes such as the addition of a single methyl group to a binding partner can sometimes suffice to nearly eliminate binding; for example, the aminoacyl tRNA synthetase specific to the amino acid valine discriminates against the very similar side chain of the amino acid isoleucine.

Proteins can bind to other proteins as well as to small-molecule substrates. When proteins bind specifically to other copies of the same molecule, they can oligomerize to form fibrils; this process occurs often in structural proteins that consist of globular monomers that self-associate to form rigid fibers. Protein–protein interactions also regulate enzymatic activity, control progression through the cell cycle, and allow the assembly of large protein complexes that carry out many closely related reactions with a common biological function. Proteins can also bind to, or even be integrated into, cell membranes. The ability of binding partners to induce conformational changes in proteins allows the construction of enormously complex signaling networks. As interactions between proteins are reversible, and depend heavily on the availability of different groups of partner proteins to form aggregates that are capable to carry out discrete sets of function, study of the interactions between specific proteins is a key to understand important aspects of cellular function, and ultimately the properties that distinguish particular cell types.

Enzymes

The best-known role of proteins in the cell is as enzymes, which catalyse chemical reactions. Enzymes are usually highly specific and accelerate only one or a few chemical reactions. Enzymes carry out most of the reactions involved in metabolism, as well as manipulating DNA in processes such as DNA replication, DNA repair, and transcription. Some enzymes act on other proteins to add or remove chemical groups in a process known as posttranslational modification. About 4,000 reactions are known to be catalysed by enzymes. The rate acceleration conferred by enzymatic catalysis is often enormous—as much as 10-fold increase in rate over the uncatalysed reaction in the case of orotate decarboxylase (78 million years without the enzyme, 18 milliseconds with the enzyme).

The molecules bound and acted upon by enzymes are called substrates. Although enzymes can consist of hundreds of amino acids, it is usually only a small fraction of the residues that come in contact with the substrate, and an even smaller fraction—three to four residues on average—that are directly involved in catalysis. The region of the enzyme that binds the substrate and contains the catalytic residues is known as the active site.

Dirigent proteins are members of a class of proteins that dictate the stereochemistry of a compound synthesized by other enzymes.

Cell Signaling and Ligand Binding

Ribbon diagram of a mouse antibody against cholera that binds a carbohydrate antigen

Many proteins are involved in the process of cell signaling and signal transduction. Some proteins, such as insulin, are extracellular proteins that transmit a signal from the cell in which they were synthesized to other cells in distant tissues. Others are membrane proteins that act as receptors whose main function is to bind a signaling molecule and induce a biochemical response in the cell. Many receptors have a binding site exposed on the cell surface and an effector domain within the cell, which may have enzymatic activity or may undergo a conformational change detected by other proteins within the cell.

Antibodies are protein components of an adaptive immune system whose main function is to bind antigens, or foreign substances in the body, and target them for destruction. Antibodies can be secreted into the extracellular environment or anchored in the membranes of specialized B cells known as plasma cells. Whereas enzymes are limited in their binding affinity for their substrates by the necessity of conducting their reaction, antibodies have no such constraints. An antibody's binding affinity to its target is extraordinarily high.

Many ligand transport proteins bind particular small biomolecules and transport them to other locations in the body of a multicellular organism. These proteins must have a high binding affinity when their ligand is present in high concentrations, but must also release the ligand when it is present at low concentrations in the target tissues. The canonical example of a ligand-binding protein is haemoglobin, which transports oxygen from the lungs to other organs and tissues in all vertebrates and has close homologs in every biological kingdom. Lectins are sugar-binding proteins which are highly specific for their sugar moieties. Lectins typically play a role in biological recognition phenomena involving cells and proteins. Receptors and hormones are highly specific binding proteins.

Transmembrane proteins can also serve as ligand transport proteins that alter the permeability of the cell membrane to small molecules and ions. The membrane alone has a hydrophobic core through which polar or charged molecules cannot diffuse. Membrane proteins contain internal channels that allow such molecules to enter and exit the cell. Many ion channel proteins are specialized to select for only a particular ion; for example, potassium and sodium channels often discriminate for only one of the two ions.

Structural Proteins

Structural proteins confer stiffness and rigidity to otherwise-fluid biological components. Most structural proteins are fibrous proteins; for example, collagen and elastin are critical components of connective tissue such as cartilage, and keratin is found in hard or filamentous structures such as hair, nails, feathers, hooves, and some animal shells. Some globular proteins can also play structural functions, for example, actin and tubulin are globular and soluble as monomers, but polymerize to form long, stiff fibers that make up the cytoskeleton, which allows the cell to maintain its shape and size.

Other proteins that serve structural functions are motor proteins such as myosin, kinesin, and dynein, which are capable of generating mechanical forces. These proteins are crucial for cellular motility of single celled organisms and the sperm of many multicellular organisms which reproduce sexually. They also generate the forces exerted by contracting muscles and play essential roles in intracellular transport.

Methods of Study

The activities and structures of proteins may be examined in vitro, in vivo, and in silico. In vitro studies of purified proteins in controlled environments are useful for learning how a protein carries out its function: for example, enzyme kinetics studies explore the chemical mechanism of an enzyme's catalytic activity and its relative affinity for various possible substrate molecules. By contrast, in vivo experiments can provide information about the physiological role of a protein in the context of a cell or even a whole organism. In silico studies use computational methods to study proteins.

Protein Purification

To perform in vitro analysis, a protein must be purified away from other cellular components. This process usually begins with cell lysis, in which a cell's membrane is disrupted and its internal contents released into a solution known as a crude lysate. The resulting mixture can be purified using ultracentrifugation, which fractionates the various cellular components into fractions containing soluble proteins; membrane lipids and proteins; cellular organelles, and nucleic acids. Precipitation by a method known as salting out can concentrate the proteins from this lysate. Various types of chromatography are then used to isolate the protein or proteins of interest based on properties such as molecular weight, net charge and binding affinity. The level of purification can be monitored using various types of gel electrophoresis if the desired protein's molecular weight and isoelectric point are known, by spectroscopy if the protein has distinguishable spectroscopic features, or by enzyme assays if the protein has enzymatic activity. Additionally, proteins can be isolated according their charge using electrofocusing.

For natural proteins, a series of purification steps may be necessary to obtain protein sufficiently pure for laboratory applications. To simplify this process, genetic engineering is often used to add chemical features to proteins that make them easier to purify without affecting their structure or activity. Here, a "tag" consisting of a specific amino acid sequence, often a series of histidine residues (a "His-tag"), is attached to one terminus of the protein. As a result, when the lysate is passed over a chromatography column containing nickel, the histidine residues ligate the nickel and attach to the column while the untagged components of the lysate pass unimpeded. A number of different tags have been developed to help researchers purify specific proteins from complex mixtures.

Cellular Localization

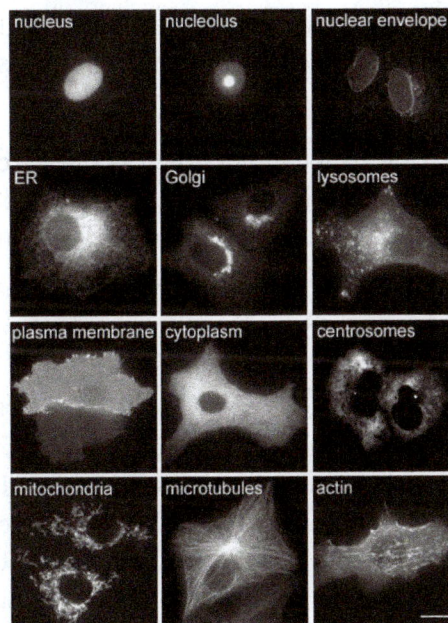

with friendly permission of Jeremy Simpson and Rainer Pepperkok

Proteins in different cellular compartments and structures tagged with green fluorescent protein (here, white)

The study of proteins in vivo is often concerned with the synthesis and localization of the protein within the cell. Although many intracellular proteins are synthesized in the cytoplasm and membrane-bound or secreted proteins in the endoplasmic reticulum, the specifics of how proteins are targeted to specific organelles or cellular structures is often unclear. A useful technique for assessing cellular localization uses genetic engineering to express in a cell a fusion protein or chimera consisting of the natural protein of interest linked to a "reporter" such as green fluorescent protein (GFP). The fused protein's position within the cell can be cleanly and efficiently visualized using microscopy, as shown in the figure opposite.

Other methods for elucidating the cellular location of proteins requires the use of known compartmental markers for regions such as the ER, the Golgi, lysosomes or vacuoles, mitochondria, chloroplasts, plasma membrane, etc. With the use of fluorescently tagged versions of these markers or of antibodies to known markers, it becomes much simpler to identify the localization of a protein of interest. For example, indirect immunofluorescence will allow for fluorescence colocalization and demonstration of location. Fluorescent dyes are used to label cellular compartments for a similar purpose.

Other possibilities exist, as well. For example, immunohistochemistry usually utilizes an antibody to one or more proteins of interest that are conjugated to enzymes yielding either luminescent or chromogenic signals that can be compared between samples, allowing for localization information. Another applicable technique is cofractionation in sucrose (or other material) gradients using isopycnic centrifugation. While this technique does not prove colocalization of a compartment of known density and the protein of interest, it does increase the likelihood, and is more amenable to large-scale studies.

Finally, the gold-standard method of cellular localization is immunoelectron microscopy. This technique also uses an antibody to the protein of interest, along with classical electron microscopy techniques. The sample is prepared for normal electron microscopic examination, and then treated with an antibody to the protein of interest that is conjugated to an extremely electro-dense material, usually gold. This allows for the localization of both ultrastructural details as well as the protein of interest.

Through another genetic engineering application known as site-directed mutagenesis, researchers can alter the protein sequence and hence its structure, cellular localization, and susceptibility to regulation. This technique even allows the incorporation of unnatural amino acids into proteins, using modified tRNAs, and may allow the rational design of new proteins with novel properties.

Proteomics

The total complement of proteins present at a time in a cell or cell type is known as its proteome, and the study of such large-scale data sets defines the field of proteomics, named by analogy to the related field of genomics. Key experimental techniques in proteomics include 2D electrophoresis, which allows the separation of a large number of proteins, mass spectrometry, which allows rapid high-throughput identification of proteins and sequencing of peptides (most often after in-gel digestion), protein microarrays, which allow the detection of the relative levels of a large number of proteins present in a cell, and two-hybrid screening, which allows the systematic exploration of protein–protein interactions. The total complement of biologically possible such interactions is known as the interactome. A systematic attempt to determine the structures of proteins representing every possible fold is known as structural genomics.

Bioinformatics

A vast array of computational methods have been developed to analyze the structure, function, and evolution of proteins.

The development of such tools has been driven by the large amount of genomic and proteomic data available for a variety of organisms, including the human genome. It is simply impossible to study all proteins experimentally, hence only a few are subjected to laboratory experiments while computational tools are used to extrapolate to similar proteins. Such homologous proteins can be efficiently identified in distantly related organisms by sequence alignment. Genome and gene sequences can be searched by a variety of tools for certain properties. Sequence profiling tools can find restriction enzyme sites, open reading frames in nucleotide sequences, and predict secondary structures. Phylogenetic trees can be constructed and evolutionary hypotheses developed using special software like ClustalW regarding the ancestry of modern organisms and the genes they express. The field of bioinformatics is now indispensable for the analysis of genes and proteins.

Structure Prediction and Simulation

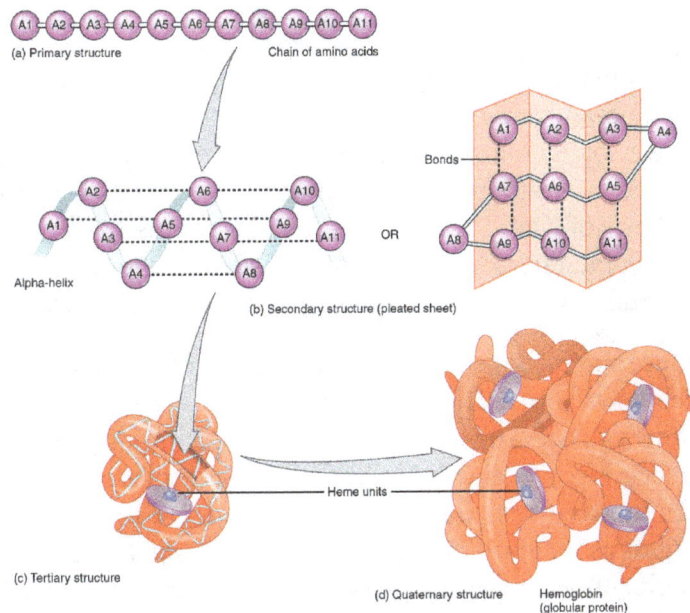

Constituent amino-acids can be analyzed to predict secondary, tertiary and quaternary protein structure, in this case hemoglobin containing heme units

Complementary to the field of structural genomics, protein structure prediction seeks to develop efficient ways to provide plausible models for proteins whose structures have not yet been determined experimentally. The most successful type of structure prediction, known as homology modeling, relies on the existence of a "template" structure with sequence similarity to the protein being modeled; structural genomics' goal is to provide sufficient representation in solved structures to model most of those that remain. Although producing accurate models remains a challenge when only distantly related template structures are available, it has been suggested that sequence alignment is the bottleneck in this process, as quite accurate models can be produced if a "perfect" sequence alignment is known. Many structure prediction methods have served to inform the emerging field of protein engineering, in which novel protein folds have already been designed. A more complex computational problem is the prediction of intermolecular interactions, such as in molecular docking and protein–protein interaction prediction.

The processes of protein folding and binding can be simulated using such technique as molecular mechanics, in particular, molecular dynamics and Monte Carlo, which increasingly take advantage of parallel and distributed computing (Folding@home project; molecular modeling on GPU). The folding of small α-helical protein domains such as the villin headpiece and the HIV accessory protein have been successfully simulated in silico, and hybrid methods that combine standard molecular dynamics with quantum mechanics calculations have allowed exploration of the electronic states of rhodopsins.

Protein Disorder and Unstructure Prediction

Many proteins (in Eucaryota ~33%) contain large unstructured but biologically functional segments and can be classified as intrinsically disordered proteins. Predicting and analysing protein disorder is, therefore, an important part of protein structure characterisation.

Nutrition

Most microorganisms and plants can biosynthesize all 20 standard amino acids, while animals (including humans) must obtain some of the amino acids from the diet. The amino acids that an organism cannot synthesize on its own are referred to as essential amino acids. Key enzymes that synthesize certain amino acids are not present in animals — such as aspartokinase, which catalyses the first step in the synthesis of lysine, methionine, and threonine from aspartate. If amino acids are present in the environment, microorganisms can conserve energy by taking up the amino acids from their surroundings and downregulating their biosynthetic pathways.

In animals, amino acids are obtained through the consumption of foods containing protein. Ingested proteins are then broken down into amino acids through digestion, which typically involves denaturation of the protein through exposure to acid and hydrolysis by enzymes called proteases. Some ingested amino acids are used for protein biosynthesis, while others are converted to glucose through gluconeogenesis, or fed into the citric acid cycle. This use of protein as a fuel is particularly important under starvation conditions as it allows the body's own proteins to be used to support life, particularly those found in muscle. Amino acids are also an important dietary source of nitrogen.

History and Etymology

Proteins were recognized as a distinct class of biological molecules in the eighteenth century by Antoine Fourcroy and others, distinguished by the molecules' ability to coagulate or flocculate under treatments with heat or acid. Noted examples at the time included albumin from egg whites, blood serum albumin, fibrin, and wheat gluten.

Proteins were first described by the Dutch chemist Gerardus Johannes Mulder and named by the Swedish chemist Jöns Jacob Berzelius in 1838. Mulder carried out elemental analysis of common proteins and found that nearly all proteins had the same empirical formula, $C_{400}H_{620}N_{100}O_{120}P_1S_1$. He came to the erroneous conclusion that they might be composed of a single type of (very large) molecule. The term "protein" to describe these molecules was proposed by Mulder's associate Berzelius.

Mulder went on to identify the products of protein degradation such as the amino acid leucine for which he found a (nearly correct) molecular weight of 131 Da.

Early nutritional scientists such as the German Carl von Voit believed that protein was the most important nutrient for maintaining the structure of the body, because it was generally believed that "flesh makes flesh." Karl Heinrich Ritthausen extended known protein forms with the identification of glutamic acid. At the Connecticut Agricultural Experiment Station a detailed review of the vegetable proteins was compiled by Thomas Burr Osborne. Working with Lafayette Mendel and applying Liebig's law of the minimum in feeding laboratory rats, the nutritionally essential amino acids were established. The work was continued and communicated by William Cumming Rose. The understanding of proteins as polypeptides came through the work of Franz Hofmeister and Hermann Emil Fischer. The central role of proteins as enzymes in living organisms was not fully appreciated until 1926, when James B. Sumner showed that the enzyme urease was in fact a protein.

The difficulty in purifying proteins in large quantities made them very difficult for early protein biochemists to study. Hence, early studies focused on proteins that could be purified in large quantities, e.g., those of blood, egg white, various toxins, and digestive/metabolic enzymes obtained from slaughterhouses. In the 1950s, the Armour Hot Dog Co. purified 1 kg of pure bovine pancreatic ribonuclease A and made it freely available to scientists; this gesture helped ribonuclease A become a major target for biochemical study for the following decades.

John Kendrew with model of myoglobin in progress

Linus Pauling is credited with the successful prediction of regular protein secondary structures based on hydrogen bonding, an idea first put forth by William Astbury in 1933. Later work by Walter Kauzmann on denaturation, based partly on previous studies by Kaj Linderstrøm-Lang, contributed an understanding of protein folding and structure mediated by hydrophobic interactions.

The first protein to be sequenced was insulin, by Frederick Sanger, in 1949. Sanger correctly determined the amino acid sequence of insulin, thus conclusively demonstrating that proteins consisted of linear polymers of amino acids rather than branched chains, colloids, or cyclols. He won the Nobel Prize for this achievement in 1958.

The first protein structures to be solved were hemoglobin and myoglobin, by Max Perutz and Sir John Cowdery Kendrew, respectively, in 1958. As of 2016, the Protein Data Bank has over 115,000 atomic-resolution structures of proteins. In more recent times, cryo-electron microscopy of large macromolecular assemblies and computational protein structure prediction of small protein domains are two methods approaching atomic resolution.

Types of Proteins

Amino Acid

Amino acids are biologically important organic compounds containing amine (-NH$_2$) and carboxylic acid (-COOH) functional groups, usually along with a side-chain (R group) specific to each amino acid. The key elements of an amino acid are carbon, hydrogen, oxygen, and nitrogen, though other elements are found in the side-chains of certain amino acids. About 500 amino acids are known (though only 20 appear in the genetic code) and can be classified in many ways. They can be classified according to the core structural functional groups' locations as alpha- (α-), beta- (β-), gamma- (γ-) or delta- (δ-) amino acids; other categories relate to polarity, pH level, and side-chain group type (aliphatic, acyclic, aromatic, containing hydroxyl or sulfur, etc.). In the form of proteins, amino acids comprise the second-largest component (water is the largest) of human muscles, cells and other tissues. Outside proteins, amino acids perform critical roles in processes such as neurotransmitter transport and biosynthesis.

In biochemistry, amino acids having both the amine and the carboxylic acid groups attached to the first (alpha-) carbon atom have particular importance. They are known as 2-, alpha-, or α-amino acids (generic formula H$_2$NCHRCOOH in most cases, where R is an organic substituent known as a "side-chain"); often the term "amino acid" is used to refer specifically to these. They include the 23 proteinogenic ("protein-building") amino acids, which combine into peptide chains ("polypeptides") to form the building-blocks of a vast array of proteins. These are all L-stereoisomers ("left-handed" isomers), although a few D-amino acids ("right-handed") occur in bacterial envelopes, as a neuro-modulator (D-serine), and in some antibiotics. Twenty of the proteinogenic amino acids are encoded directly by triplet codons in the genetic code and are known as "standard" amino acids. The other three ("non-standard" or "non-canonical") are selenocysteine (present in many noneukaryotes as well as most eukaryotes, but not coded directly by DNA), pyrrolysine (found only in some archea and one bacterium) and N-formylmethionine (which is often the initial amino acid of proteins in bacteria, mitochondria, and chloroplasts). Pyrrolysine and selenocysteine are encoded via variant codons; for example, selenocysteine is encoded by stop codon and SECIS element. Codon–tRNA combinations not found in nature can also be used to "expand" the genetic code and create novel proteins known as alloproteins incorporating non-proteinogenic amino acids.

Many important proteinogenic and non-proteinogenic amino acids also play critical non-protein roles within the body. For example, in the human brain, glutamate (standard glutamic acid) and gamma-amino-butyric acid ("GABA", non-standard gamma-amino acid) are, respectively, the main excitatory and inhibitory neurotransmitters; hydroxyproline (a major component of the connective tissue collagen) is synthesised from proline; the standard amino acid glycine is used to synthesise porphyrins used in red blood cells; and the non-standard carnitine is used in lipid transport.

Nine proteinogenic amino acids are called "essential" for humans because they cannot be created from other compounds by the human body and so must be taken in as food. Others may be conditionally essential for certain ages or medical conditions. Essential amino acids may also differ between species.

Because of their biological significance, amino acids are important in nutrition and are commonly used in nutritional supplements, fertilizers, and food technology. Industrial uses include the production of drugs, biodegradable plastics, and chiral catalysts.

History

The first few amino acids were discovered in the early 19th century. In 1806, French chemists Louis-Nicolas Vauquelin and Pierre Jean Robiquet isolated a compound in asparagus that was subsequently named asparagine, the first amino acid to be discovered. Cystine was discovered in 1810, although its monomer, cysteine, remained undiscovered until 1884. Glycine and leucine were discovered in 1820. The last of the 20 common amino acids to be discovered was threonine in 1935 by William Cumming Rose, who also determined the essential amino acids and established the minimum daily requirements of all amino acids for optimal growth.

Usage of the term amino acid in the English language is from 1898. Proteins were found to yield amino acids after enzymatic digestion or acid hydrolysis. In 1902, Emil Fischer and Franz Hofmeister proposed that proteins are the result of the formation of bonds between the amino group of one amino acid with the carboxyl group of another, in a linear structure that Fischer termed "peptide".

General Structure

In the structure shown at the top of the page, R represents a side-chain specific to each amino acid. The carbon atom next to the carboxyl group (which is therefore numbered 2 in the carbon chain starting from that functional group) is called the α−carbon. Amino acids containing an amino group bonded directly to the alpha carbon are referred to as alpha amino acids. These include amino acids such as proline which contain secondary amines, which used to be often referred to as "imino acids".

Isomerism

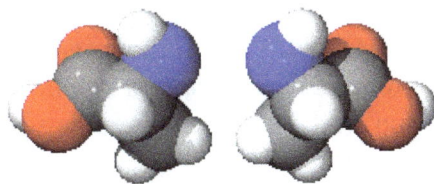

The two enantiomers of alanine, D-alanine and L-alanine

The alpha amino acids are the most common form found in nature, but only when occurring in the L-isomer. The alpha carbon is a chiral carbon atom, with the exception of glycine which has two indistinguishable hydrogen atoms on the alpha carbon. Therefore, all alpha amino acids but glycine can exist in either of two enantiomers, called L or D amino acids, which are mirror images of each other. While L-amino acids represent all of the amino acids found in proteins during translation in the ribosome, D-amino acids are found in some proteins produced by enzyme posttranslational modifications after translation and translocation to the endoplasmic reticulum, as in exotic sea-dwelling organisms such as cone snails. They are also abundant components of the peptidoglycan cell walls of bacteria, and D-serine may act as a neurotransmitter in the brain. D-amino acids are used in racemic crystallography to create centrosymmetric crystals, which (depending on the protein) may allow for easier and more robust protein structure determination. The L and D convention for amino acid configuration refers not to the optical activity of the amino acid itself but rather to the optical activity of the isomer of glyceraldehyde from which that amino acid can, in theory, be synthesized (D-glyceraldehyde is dextrorotatory; L-glyceraldehyde is levorotatory). In alternative fashion, the (S) and (R) designators are used to indicate the absolute stereochemistry. Almost all of the amino acids in proteins are (S) at the α carbon, with cysteine being (R) and glycine non-chiral. Cysteine has its side-chain in the same geometric position as the other amino acids, but the R/S terminology is reversed because of the higher atomic number of sulfur compared to the carboxyl oxygen gives the side-chain a higher priority, whereas the atoms in most other side-chains give them lower priority.

Side Chains

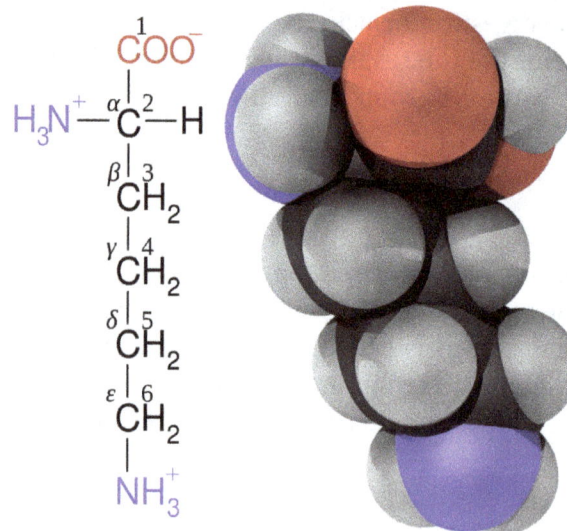

Lysine with the carbon atoms in the side-chain labeled

In amino acids that have a carbon chain attached to the α−carbon (such as lysine, shown to the right) the carbons are labeled in order as α, β, γ, δ, and so on. In some amino acids, the amine group is attached to the β or γ-carbon, and these are therefore referred to as beta or gamma amino acids.

Amino acids are usually classified by the properties of their side-chain into four groups. The side-chain can make an amino acid a weak acid or a weak base, and a hydrophile if the side-chain is polar or a hydrophobe if it is nonpolar. The chemical structures of the 22 standard amino acids, along with their chemical properties, are described more fully in the article on these proteinogenic amino acids.

The phrase "branched-chain amino acids" or BCAA refers to the amino acids having aliphatic side-chains that are non-linear; these are leucine, isoleucine, and valine. Proline is the only proteinogenic amino acid whose side-group links to the α-amino group and, thus, is also the only pro-teinogenic amino acid containing a secondary amine at this position. In chemical terms, proline is, therefore, an imino acid, since it lacks a primary amino group, although it is still classed as an amino acid in the current biochemical nomenclature, and may also be called an "N-alkylated alpha-amino acid".

Zwitterions

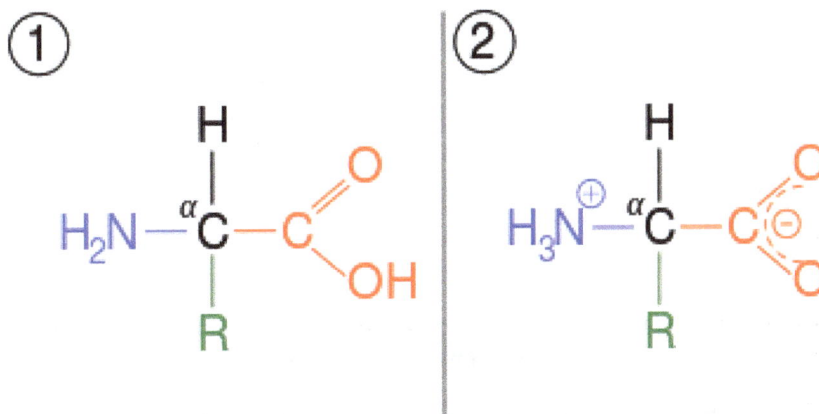

An amino acid in its (1) un-ionized and (2) zwitterionic forms

The α-carboxylic acid group of amino acids is a weak acid, meaning that it releases a hydron (such as a proton) at moderate pH values. In other words, carboxylic acid groups ($-CO_2H$) can be de-protonated to become negative carboxylates ($-CO_2^-$). The negatively charged carboxylate ion predominates at pH values greater than the pKa of the carboxylic acid group (mean for the 20 common amino acids is about 2.2). In a complementary fashion, the α-amine of amino acids is a weak base, meaning that it accepts a proton at moderate pH values. In other words, α-amino groups (NH_2-) can be protonated to become positive α-ammonium groups ($^+NH_3-$). The positive-ly charged α-ammonium group predominates at pH values less than the pKa of the α-ammonium group (mean for the 20 common α-amino acids is about 9.4).

Because all amino acids contain amine and carboxylic acid functional groups, they share amphi-protic properties. Below pH 2.2, the predominant form will have a neutral carboxylic acid group and a positive α-ammonium ion (net charge +1), and above pH 9.4, a negative carboxylate and neutral α-amino group (net charge −1). But at pH between 2.2 and 9.4, an amino acid usually con-tains both a negative carboxylate and a positive α-ammonium group, as shown in structure (2) on the right, so has net zero charge. This molecular state is known as a zwitterion, from the German Zwitter meaning hermaphrodite or hybrid. The fully neutral form (structure (1) on the right) is a very minor species in aqueous solution throughout the pH range (less than 1 part in 10). Amino acids exist as zwitterions also in the solid phase, and crystallize with salt-like properties unlike typical organic acids or amines.

Isoelectric Point

The variation in titration curves when the amino acids are grouped by category can be seen here. With the exception of tyrosine, using titration to differentiate between hydrophobic amino acids is problematic.

Titration Curves of 20 Amino Acids

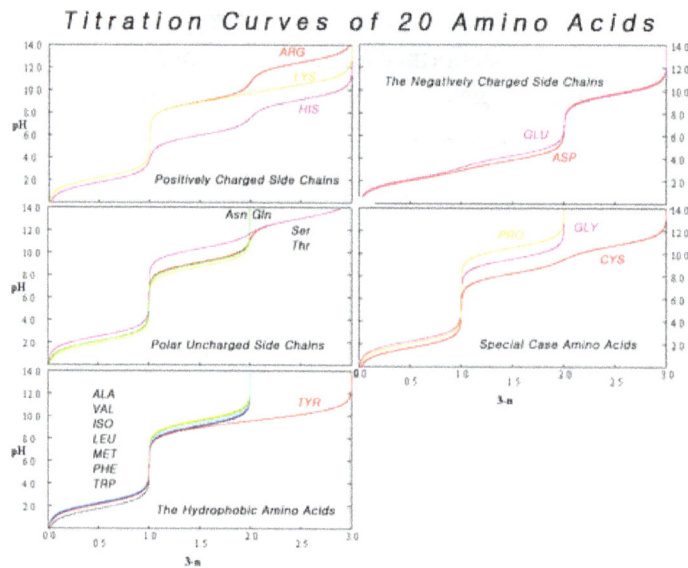

Composite of Titration Curves Grouped by Side Chain Category using applet http://cti.itc.virginia.edu/~cmg/Demo/compareAA/compareAAApplet.html

At pH values between the two pKa values, the zwitterion predominates, but coexists in dynamic equilibrium with small amounts of net negative and net positive ions. At the exact midpoint between the two pKa values, the trace amount of net negative and trace of net positive ions exactly balance, so that average net charge of all forms present is zero. This pH is known as the isoelectric point pI, so pI = $\frac{1}{2}$(pKa$_1$ + pKa$_2$). The individual amino acids all have slightly different pKa values, so have different isoelectric points. For amino acids with charged side-chains, the pKa of the side-chain is involved. Thus for Asp, Glu with negative side-chains, pI = $\frac{1}{2}$(pKa$_1$ + pKa$_R$), where pKa$_R$ is the side-chain pKa. Cysteine also has potentially negative side-chain with pKa$_R$ = 8.14, so pI should be calculated as for Asp and Glu, even though the side-chain is not significantly charged at neutral pH. For His, Lys, and Arg with positive side-chains, pI = $\frac{1}{2}$(pKa$_R$ + pKa$_2$). Amino acids have zero mobility in electrophoresis at their isoelectric point, although this behaviour is more usually exploited for peptides and proteins than single amino acids. Zwitterions have minimum solubility at their isoelectric point and some amino acids (in particular, with non-polar side-chains) can be isolated by precipitation from water by adjusting the pH to the required isoelectric point.

Occurrence and Functions in Biochemistry

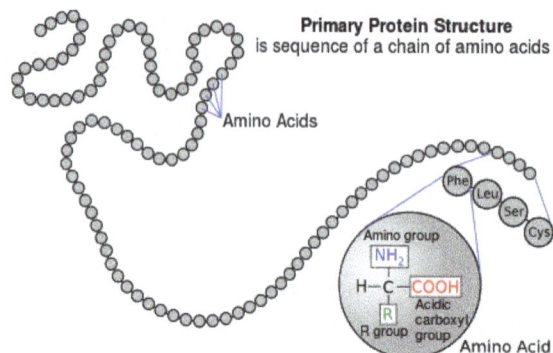

A polypeptide is an unbranched chain of amino acids.

Proteinogenic Amino Acids

Amino acids are the structural units (monomers) that make up proteins. They join together to form short polymer chains called peptides or longer chains called either polypeptides or proteins. These polymers are linear and unbranched, with each amino acid within the chain attached to two neighboring amino acids. The process of making proteins is called translation and involves the step-by-step addition of amino acids to a growing protein chain by a ribozyme that is called a ribosome. The order in which the amino acids are added is read through the genetic code from an mRNA template, which is a RNA copy of one of the organism's genes.

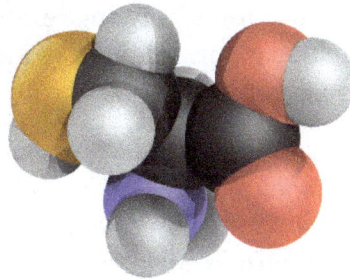

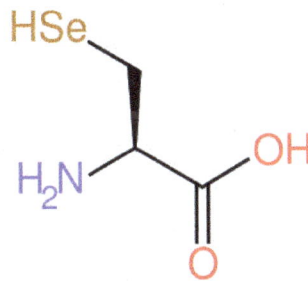

The amino acid selenocysteine

Twenty-two amino acids are naturally incorporated into polypeptides and are called proteinogenic or natural amino acids. Of these, 20 are encoded by the universal genetic code. The remaining 2, selenocysteine and pyrrolysine, are incorporated into proteins by unique synthetic mechanisms. Selenocysteine is incorporated when the mRNA being translated includes a SECIS element, which causes the UGA codon to encode selenocysteine instead of a stop codon. Pyrrolysine is used by some methanogenic archaea in enzymes that they use to produce methane. It is coded for with the codon UAG, which is normally a stop codon in other organisms. This UAG codon is followed by a PYLIS downstream sequence.

Non-Proteinogenic Amino Acids

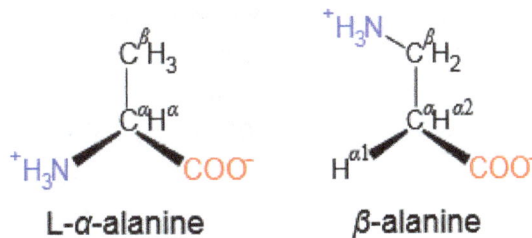

β-alanine and its α-alanine isomer

Aside from the 22 proteinogenic amino acids, there are many other amino acids that are called non-proteinogenic. Those either are not found in proteins (for example carnitine, GABA) or are not produced directly and in isolation by standard cellular machinery (for example, hydroxyproline and selenomethionine).

Non-proteinogenic amino acids that are found in proteins are formed by post-translational modification, which is modification after translation during protein synthesis. These modifications are often essential for the function or regulation of a protein; for example, the carboxylation of glutamate allows for better binding of calcium cations, and the hydroxylation of proline is critical for maintaining connective tissues. Another example is the formation of hypusine in the translation initiation factor EIF5A, through modification of a lysine residue. Such modifications can also determine the localization of the protein, e.g., the addition of long hydrophobic groups can cause a protein to bind to a phospholipid membrane.

Some non-proteinogenic amino acids are not found in proteins. Examples include lanthionine, 2-aminoisobutyric acid, dehydroalanine, and the neurotransmitter gamma-aminobutyric acid. Non-proteinogenic amino acids often occur as intermediates in the metabolic pathways for standard amino acids – for example, ornithine and citrulline occur in the urea cycle, part of amino acid catabolism. A rare exception to the dominance of α-amino acids in biology is the β-amino acid beta alanine (3-aminopropanoic acid), which is used in plants and microorganisms in the synthesis of pantothenic acid (vitamin B$_5$), a component of coenzyme A.

D-Amino Acid Natural Abundance

D-isomers are uncommon in live organisms. For instance, gramicidin is a polypeptide made up from mixture of D- and L-amino acids. Other compounds containing D-amino acids are tyrocidine and valinomycin. These compounds disrupt bacterial cell walls, particularly in Gram-positive bacteria. Only 837 D-amino acids were found in Swiss-Prot database (187 million amino acids analysed).

Non-Standard Amino Acids

The 20 amino acids that are encoded directly by the codons of the universal genetic code are called standard or canonical amino acids. The others are called non-standard or non-canonical. Most of the non-standard amino acids are also non-proteinogenic (i.e. they cannot be used to build proteins), but three of them are proteinogenic, as they can be used to build proteins by exploiting information not encoded in the universal genetic code.

The three non-standard proteinogenic amino acids are selenocysteine (present in many non-eukaryotes as well as most eukaryotes, but not coded directly by DNA), pyrrolysine (found only in some archaea and one bacterium), and N-formylmethionine (which is often the initial amino acid of proteins in bacteria, mitochondria, and chloroplasts). For example, 25 human proteins include selenocysteine (Sec) in their primary structure, and the structurally characterized enzymes (selenoenzymes) employ Sec as the catalytic moiety in their active sites. Pyrrolysine and selenocysteine are encoded via variant codons. For example, selenocysteine is encoded by stop codon and SECIS element.

In Human Nutrition

When taken up into the human body from the diet, the 20 standard amino acids either are used to synthesize proteins and other biomolecules or are oxidized to urea and carbon dioxide as a source of energy. The oxidation pathway starts with the removal of the amino group by a transaminase; the amino group is then fed into the urea cycle. The other product of transamidation is a keto acid that enters the citric acid cycle. Glucogenic amino acids can also be converted into glucose, through gluconeogenesis. Of the 20 standard amino acids, nine (His, Ile, Leu, Lys, Met, Phe, Thr, Trp and Val), are called essential amino acids because the human body cannot synthesize them from other compounds at the level needed for normal growth, so they must be obtained from food. In addition, cysteine, taurine, tyrosine, and arginine are considered semiessential amino-acids in children (though taurine is not technically an amino acid), because the metabolic pathways that synthesize these amino acids are not fully developed. The amounts required also depend on the age and health of the individual, so it is hard to make general statements about the dietary requirement for some amino acids. Dietary exposure to the non-standard amino acid BMAA has been linked to human neurodegenerative diseases, including ALS.

Non-Protein Functions

Human biosynthesis pathway for trace amines and catecholamines

In humans, non-protein amino acids also have important roles as metabolic intermediates, such as in the biosynthesis of the neurotransmitter gamma-amino-butyric acid (GABA). Many amino acids are used to synthesize other molecules, for example:

- Tryptophan is a precursor of the neurotransmitter serotonin.

- Tyrosine (and its precursor phenylalanine) are precursors of the catecholamine neurotransmitters dopamine, epinephrine and norepinephrine and various trace amines.

- Phenylalanine is a precursor of phenethylamine and tyrosine in humans. In plants, it is a precursor of various phenylpropanoids, which are important in plant metabolism.

- Glycine is a precursor of porphyrins such as heme.

- Arginine is a precursor of nitric oxide.

- Ornithine and S-adenosylmethionine are precursors of polyamines.

- Aspartate, glycine, and glutamine are precursors of nucleotides.

However, not all of the functions of other abundant non-standard amino acids are known.

Some non-standard amino acids are used as defenses against herbivores in plants. For example, canavanine is an analogue of arginine that is found in many legumes, and in particularly large amounts in Canavalia gladiata (sword bean). This amino acid protects the plants from predators such as insects and can cause illness in people if some types of legumes are eaten without processing. The non-protein amino acid mimosine is found in other species of legume, in particular Leucaena leucocephala. This compound is an analogue of tyrosine and can poison animals that graze on these plants.

Uses in Industry

Amino acids are used for a variety of applications in industry, but their main use is as additives to animal feed. This is necessary, since many of the bulk components of these feeds, such as soybeans, either have low levels or lack some of the essential amino acids: lysine, methionine, threonine, and tryptophan are most important in the production of these feeds. In this industry, amino acids are also used to chelate metal cations in order to improve the absorption of minerals from supplements, which may be required to improve the health or production of these animals.

The food industry is also a major consumer of amino acids, in particular, glutamic acid, which is used as a flavor enhancer, and aspartame (aspartyl-phenylalanine-1-methyl ester) as a low-calorie artificial sweetener. Similar technology to that used for animal nutrition is employed in the human nutrition industry to alleviate symptoms of mineral deficiencies, such as anemia, by improving mineral absorption and reducing negative side effects from inorganic mineral supplementation.

The chelating ability of amino acids has been used in fertilizers for agriculture to facilitate the delivery of minerals to plants in order to correct mineral deficiencies, such as iron chlorosis. These fertilizers are also used to prevent deficiencies from occurring and improving the overall health of the plants. The remaining production of amino acids is used in the synthesis of drugs and cosmetics.

Similarly, some amino acids derivatives are used in pharmaceutical industry. They include 5-HTP (5-hydroxytryptophan) used for experimental treatment of depression, L-DOPA (L-dihydroxyphenylalanine) for Parkinson's treatment, and eflornithine drug that inhibits ornithine decarboxylase and used in the treatment of sleeping sickness.

Expanded Genetic Code

Since 2001, 40 non-natural amino acids have been added into protein by creating a unique codon (recoding) and a corresponding transfer-RNA:aminoacyl – tRNA-synthetase pair to encode it with diverse physicochemical and biological properties in order to be used as a tool to exploring protein structure and function or to create novel or enhanced proteins.

Nullomers

Nullomers are codons that in theory code for an amino acid, however in nature there is a selective bias against using this codon in favor of another, for example bacteria prefer to use CGA instead of AGA to code for arginine. This creates some sequences that do not appear in the genome. This characteristic can be taken advantage of and used to create new selective cancer-fighting drugs and to prevent cross-contamination of DNA samples from crime-scene investigations.

Chemical Building Blocks

Amino acids are important as low-cost feedstocks. These compounds are used in chiral pool synthesis as enantiomerically pure building-blocks.

Amino acids have been investigated as precursors chiral catalysts, e.g., for asymmetric hydrogenation reactions, although no commercial applications exist.

Biodegradable Plastics

Amino acids are under development as components of a range of biodegradable polymers. These materials have applications as environmentally friendly packaging and in medicine in drug delivery and the construction of prosthetic implants. These polymers include polypeptides, polyamides, poly-esters, polysulfides, and polyurethanes with amino acids either forming part of their main chains or bonded as side-chains. These modifications alter the physical properties and reactivities of the polymers. An interesting example of such materials is polyaspartate, a water-soluble biodegradable polymer that may have applications in disposable diapers and agriculture. Due to its solubility and ability to chelate metal ions, polyaspartate is also being used as a biodegradeable anti-scaling agent and a corrosion inhibitor. In addition, the aromatic amino acid tyrosine is being developed as a pos-sible replacement for toxic phenols such as bisphenol A in the manufacture of polycarbonates.

Reactions

As amino acids have both a primary amine group and a primary carboxyl group, these chemicals can undergo most of the reactions associated with these functional groups. These include nucleo-philic addition, amide bond formation, and imine formation for the amine group, and esterifica-tion, amide bond formation, and decarboxylation for the carboxylic acid group. The combination of these functional groups allow amino acids to be effective polydentate ligands for metal-amino acid chelates. The multiple side-chains of amino acids can also undergo chemical reactions. The types of these reactions are determined by the groups on these side-chains and are, therefore, dif-ferent between the various types of amino acid.

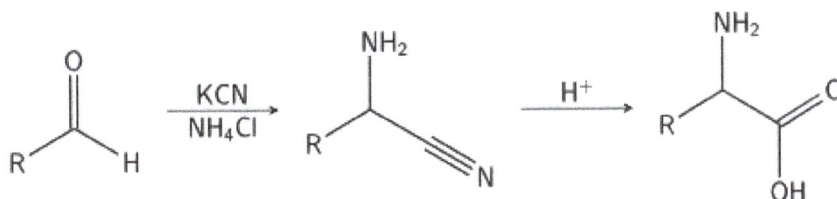

The Strecker amino acid synthesis

Chemical Synthesis

Several methods exist to synthesize amino acids. One of the oldest methods begins with the bro-mination at the α-carbon of a carboxylic acid. Nucleophilic substitution with ammonia then con-verts the alkyl bromide to the amino acid. In alternative fashion, the Strecker amino acid synthesis involves the treatment of an aldehyde with potassium cyanide and ammonia, this produces an α-amino nitrile as an intermediate. Hydrolysis of the nitrile in acid then yields a α-amino acid. Us-ing ammonia or ammonium salts in this reaction gives unsubstituted amino acids, whereas substi-tuting primary and secondary amines will yield substituted amino acids. Likewise, using ketones, instead of aldehydes, gives α,α-disubstituted amino acids. The classical synthesis gives racemic mixtures of α-amino acids as products, but several alternative procedures using asymmetric aux-iliaries or asymmetric catalysts have been developed.

At the current time, the most-adopted method is an automated synthesis on a solid support (e.g., polystyrene beads), using protecting groups (e.g., Fmoc and t-Boc) and activating groups (e.g., DCC and DIC).

Peptide Bond Formation

As both the amine and carboxylic acid groups of amino acids can react to form amide bonds, one amino acid molecule can react with another and become joined through an amide linkage. This polymerization of amino acids is what creates proteins. This condensation reaction yields the newly formed peptide bond and a molecule of water. In cells, this reaction does not occur directly; instead, the amino acid is first activated by attachment to a transfer RNA molecule through an ester bond. This aminoacyl-tRNA is produced in an ATP-dependent reaction carried out by an aminoacyl tRNA synthetase. This aminoacyl-tRNA is then a substrate for the ribosome, which catalyzes the attack of the amino group of the elongating protein chain on the ester bond. As a result of this mechanism, all proteins made by ribosomes are synthesized starting at their N-terminus and moving toward their C-terminus.

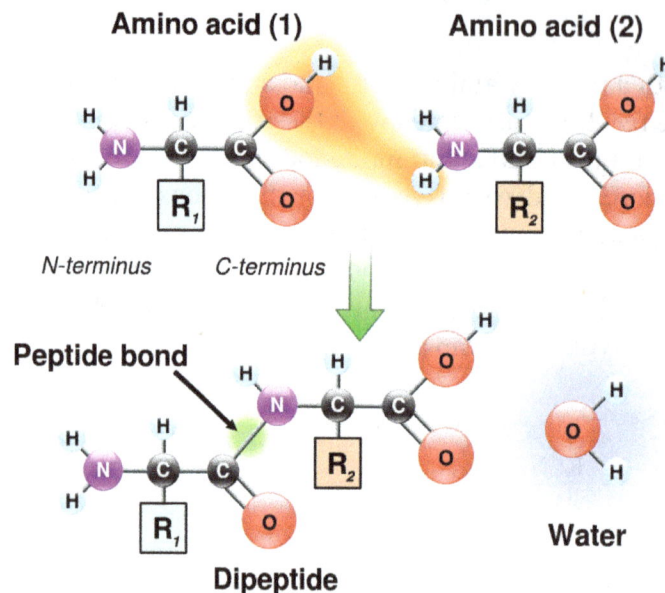

The condensation of two amino acids to form a dipeptide through a peptide bond

However, not all peptide bonds are formed in this way. In a few cases, peptides are synthesized by specific enzymes. For example, the tripeptide glutathione is an essential part of the defenses of cells against oxidative stress. This peptide is synthesized in two steps from free amino acids. In the first step, gamma-glutamylcysteine synthetase condenses cysteine and glutamic acid through a peptide bond formed between the side-chain carboxyl of the glutamate (the gamma carbon of this side-chain) and the amino group of the cysteine. This dipeptide is then condensed with glycine by glutathione synthetase to form glutathione.

In chemistry, peptides are synthesized by a variety of reactions. One of the most-used in solid-phase peptide synthesis uses the aromatic oxime derivatives of amino acids as activated units. These are added in sequence onto the growing peptide chain, which is attached to a solid resin support. The ability to easily synthesize vast numbers of different peptides by varying the types and order of amino acids (using combinatorial chemistry) has made peptide synthesis particularly important in creating libraries of peptides for use in drug discovery through high-throughput screening.

Biosynthesis

In plants, nitrogen is first assimilated into organic compounds in the form of glutamate, formed from alpha-ketoglutarate and ammonia in the mitochondrion. In order to form other amino acids, the plant uses transaminases to move the amino group to another alpha-keto carboxylic acid. For example, aspartate aminotransferase converts glutamate and oxaloacetate to alpha-ketoglutarate and aspartate. Other organisms use transaminases for amino acid synthesis, too.

Nonstandard amino acids are usually formed through modifications to standard amino acids. For example, homocysteine is formed through the transsulfuration pathway or by the demethylation of methionine via the intermediate metabolite S-adenosyl methionine, while hydroxyproline is made by a posttranslational modification of proline.

Microorganisms and plants can synthesize many uncommon amino acids. For example, some microbes make 2-aminoisobutyric acid and lanthionine, which is a sulfide-bridged derivative of alanine. Both of these amino acids are found in peptidic lantibiotics such as alamethicin. However, in plants, 1-aminocyclopropane-1-carboxylic acid is a small disubstituted cyclic amino acid that is a key intermediate in the production of the plant hormone ethylene.

Catabolism

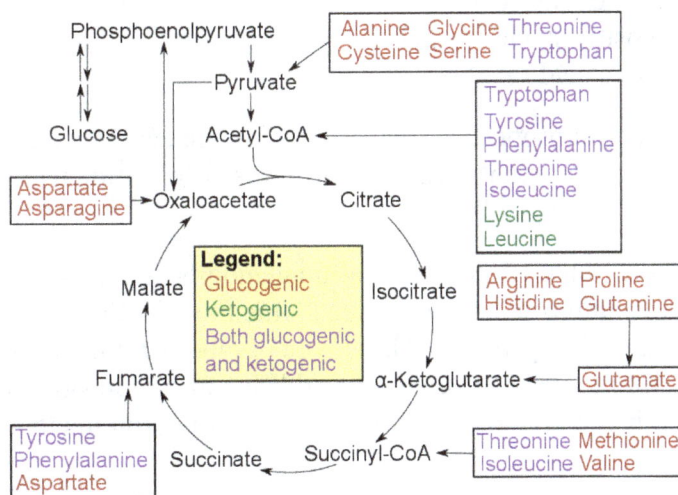

Catabolism of proteinogenic amino acids. Amino acids can be classified according to the properties of their main products as either of the following:
* Glucogenic, with the products having the ability to form glucose by gluconeogenesis
* Ketogenic, with the products not having the ability to form glucose. These products may still be used for ketogenesis or lipid synthesis.
* Amino acids catabolized into both glucogenic and ketogenic products.

Amino acids must first pass out of organelles and cells into blood circulation via amino acid transporters, since the amine and carboxylic acid groups are typically ionized. Degradation of an amino acid, occurring in the liver and kidneys, often involves deamination by moving its amino group to alpha-ketoglutarate, forming glutamate. This process involves transaminases, often the same as those used in amination during synthesis. In many vertebrates, the amino group is then removed through the urea cycle and is excreted in the form of urea. However, amino acid degradation can produce uric acid or ammonia instead. For example, serine dehydratase converts serine to pyruvate and ammonia. After removal of one or more amino groups, the remainder of the molecule can sometimes be used to synthesize new amino acids, or it can be used for energy by entering glycolysis or the citric acid cycle, as detailed in image at right.t

Physicochemical Properties of Amino Acids

The 20 amino acids encoded directly by the genetic code can be divided into several groups based on their properties. Important factors are charge, hydrophilicity or hydrophobicity, size, and functional groups. These properties are important for protein structure and protein–protein interactions. The water-soluble proteins tend to have their hydrophobic residues (Leu, Ile, Val, Phe, and Trp) buried in the middle of the protein, whereas hydrophilic side-chains are exposed to the aqueous solvent. (Note that in biochemistry, a residue refers to a specific monomer within the polymeric chain of a polysaccharide, protein or nucleic acid.) The integral membrane proteins tend to have outer rings of exposed hydrophobic amino acids that anchor them into the lipid bilayer. In the case part-way between these two extremes, some peripheral membrane proteins have a patch of hydrophobic amino acids on their surface that locks onto the membrane. In similar fashion, proteins that have to bind to positively charged molecules have surfaces rich with negatively charged amino acids like glutamate and aspartate, while proteins binding to negatively charged molecules have surfaces rich with positively charged chains like lysine and arginine. There are different hydrophobicity scales of amino acid residues.

Some amino acids have special properties such as cysteine, that can form covalent disulfide bonds to other cysteine residues, proline that forms a cycle to the polypeptide backbone, and glycine that is more flexible than other amino acids.

Many proteins undergo a range of posttranslational modifications, when additional chemical groups are attached to the amino acids in proteins. Some modifications can produce hydrophobic lipoproteins, or hydrophilic glycoproteins. These type of modification allow the reversible targeting of a protein to a membrane. For example, the addition and removal of the fatty acid palmitic acid to cysteine residues in some signaling proteins causes the proteins to attach and then detach from cell membranes.

Table of Standard Amino Acid Abbreviations and Properties

Amino Acid	3-Letter	1-Letter	Side-chain class	Side-chain polarity	Side-chain charge (pH 7.4)	Hydropathy index	Absorbance λ_{max}(nm)	ε at λ_{max} (mM$^-$ cm$^-$)	MW(Weight)
Alanine	Ala	A	aliphatic	nonpolar	neutral	1.8			89.094
Arginine	Arg	R	basic	basic polar	positive	−4.5			174.203
Asparagine	Asn	N	acid (amide)	polar	neutral	−3.5			132.119
Aspartic acid	Asp	D	acid (amide)	acidic polar	negative	−3.5			133.104
Cysteine	Cys	C	sulfur-containing	nonpolar	neutral	2.5	250	0.3	121.154

Glu-tamic acid	Glu	E	acid (am-ide)	acidic polar	negative	−3.5			147.131
Gluta-mine	Gln	Q	acid (am-ide)	polar	neutral	−3.5			146.146
Glycine	Gly	G	aliphatic	nonpolar	neutral	−0.4			75.067
Histi-dine	His	H	basic	basic polar	posi-tive(10%) neu-tral(90%)	−3.2	211	5.9	155.156
Isoleu-cine	Ile	I	aliphatic	nonpolar	neutral	4.5			131.175
Leu-cine	Leu	L	aliphatic	nonpolar	neutral	3.8			131.175
Lysine	Lys	K	basic	basic polar	positive	−3.9			146.189
Methi-onine	Met	M	sulfur-containing	nonpolar	neutral	1.9			149.208
Phe-nylala-nine	Phe	F	aromatic	nonpolar	neutral	2.8	257, 206, 188	0.2, 9.3, 60.0	165.192
Proline	Pro	P	cyclic	nonpolar	neutral	−1.6			115.132
Serine	Ser	S	hydroxyl-containing	polar	neutral	−0.8			105.093
Threo-nine	Thr	T	hydroxyl-containing	polar	neutral	−0.7			119.12
Trypto-phan	Trp	W	aromatic	nonpolar	neutral	−0.9	280, 219	5.6, 47.0	204.228
Tyro-sine	Tyr	Y	aromatic	polar	neutral	−1.3	274, 222, 193	1.4, 8.0, 48.0	181.191
Valine	Val	V	aliphatic	nonpolar	neutral	4.2			117.148

Two additional amino acids are in some species coded for by codons that are usually interpreted as stop codons:

21st and 22nd amino acids	3-Let-ter	1-Letter	MW(Weight)
Selenocysteine	Sec	U	159.065
Pyrrolysine	Pyl	O	273.325

In addition to the specific amino acid codes, placeholders are used in cases where chemical or crystallographic analysis of a peptide or protein cannot conclusively determine the identity of a residue. They are also used to summarise conserved protein sequence motifs. The use of single letters to indicate sets of similar residues is similar to the use of abbreviation codes for degenerate bases.

Ambiguous Amino Acids	3-Letter	1-Letter	
Any / unknown	Xaa	X	All
Asparagine or aspartic acid	Asx	B	D, N
Glutamine or glutamic acid	Glx	Z	E, Q
Leucine or Isoleucine	Xle	J	I, L
Hydrophobic		Φ	V, I, L, F, W, Y, M
Aromatic		Ω	F, W, Y
Aliphatic		Ψ	V, I, L, M
Small		π	P, G, A, S
Hydrophilic		ζ	S, T, H, N, Q, E, D, K, R
Positively charged		+	K, R
Negatively charged		–	D, E

Unk is sometimes used instead of Xaa, but is less standard.

In addition, many non-standard amino acids have a specific code. For example, several peptide drugs, such as Bortezomib and MG132, are artificially synthesized and retain their protecting groups, which have specific codes. Bortezomib is Pyz-Phe-boroLeu, and MG132 is Z-Leu-Leu-Leu-al. To aid in the analysis of protein structure, photo-reactive amino acid analogs are available. These include photoleucine (pLeu) and photomethionine (pMet).

PEPTIDE

Peptides are biologically occurring short chains of amino acid monomers linked by peptide (amide) bonds.

The covalent chemical bonds are formed when the carboxyl group of one amino acid reacts with the amine group of another. The shortest peptides are dipeptides, consisting of 2 amino acids joined by a single peptide bond, followed by tripeptides, tetrapeptides, etc. A polypeptide is a long, continuous, and unbranched peptide chain. Hence, peptides fall under the broad chemical classes of biological oligomers and polymers, alongside nucleic acids, oligosaccharides and polysaccharides, etc.

Peptides are distinguished from proteins on the basis of size, and as an arbitrary benchmark can be understood to contain approximately 50 or fewer amino acids. Proteins consist of one or more polypeptides arranged in a biologically functional way, often bound to ligands such as coenzymes and cofactors, or to another protein or other macromolecule (DNA, RNA, etc.), or to complex macromolecular assemblies. Finally, while aspects of the lab techniques applied to peptides versus polypeptides and proteins differ (e.g., the specifics of electrophoresis, chromatography, etc.), the size boundaries that distinguish peptides from polypeptides and proteins are not absolute: long peptides such as amyloid beta have been referred to as proteins, and smaller proteins like insulin have been considered peptides.

Amino acids that have been incorporated into peptides are termed "residues" due to the release of either a hydrogen ion from the amine end or a hydroxyl ion from the carboxyl end, or both, as a water molecule is released during formation of each amide bond. All peptides except cyclic peptides have an N-terminal and C-terminal residue at the end of the peptide.

Peptide Classes

Peptides are divided into several classes, depending on how they are produced:

Milk Peptides

Two naturally occurring milk peptides are formed from the milk protein casein when digestive enzymes break this down; they can also arise from the proteinases formed by lactobacilli during the fermentation of milk.

Ribosomal Peptides

Ribosomal peptides are synthesized by translation of mRNA. They are often subjected to proteolysis to generate the mature form. These function, typically in higher organisms, as hormones and signaling molecules. Some organisms produce peptides as antibiotics, such as microcins. Since they are translated, the amino acid residues involved are restricted to those utilized by the ribosome.

However, these peptides frequently have posttranslational modifications... such as phosphorylation, hydroxylation, sulfonation, palmitoylation, glycosylation and disulfide formation. In general, they are linear, although lariat structures have been observed. More exotic manipulations do occur, such as racemization of L-amino acids to D-amino acids in platypus venom.

Nonribosomal Peptides

Nonribosomal peptides are assembled by enzymes that are specific to each peptide, rather than by the ribosome. The most common non-ribosomal peptide is glutathione, which is a component of the antioxidant defenses of most aerobic organisms. Other nonribosomal peptides are most common in unicellular organisms, plants, and fungi and are synthesized by modular enzyme complexes called nonribosomal peptide synthetases.

These complexes are often laid out in a similar fashion, and they can contain many different modules to perform a diverse set of chemical manipulations on the developing product. These peptides are often cyclic and can have highly complex cyclic structures, although linear nonribosomal peptides are also common. Since the system is closely related to the machinery for building fatty acids and polyketides, hybrid compounds are often found. The presence of oxazoles or thiazoles often indicates that the compound was synthesized in this fashion.

Peptones

Peptones are derived from animal milk or meat digested by proteolysis. In addition to containing small peptides, the resulting spray-dried material includes fats, metals, salts, vitamins and many other biological compounds. Peptones are used in nutrient media for growing bacteria and fungi.

Peptide Fragments

Peptide fragments refer to fragments of proteins that are used to identify or quantify the source protein. Often these are the products of enzymatic degradation performed in the laboratory on a controlled sample, but can also be forensic or paleontological samples that have been degraded by natural effects.

Peptide Synthesis

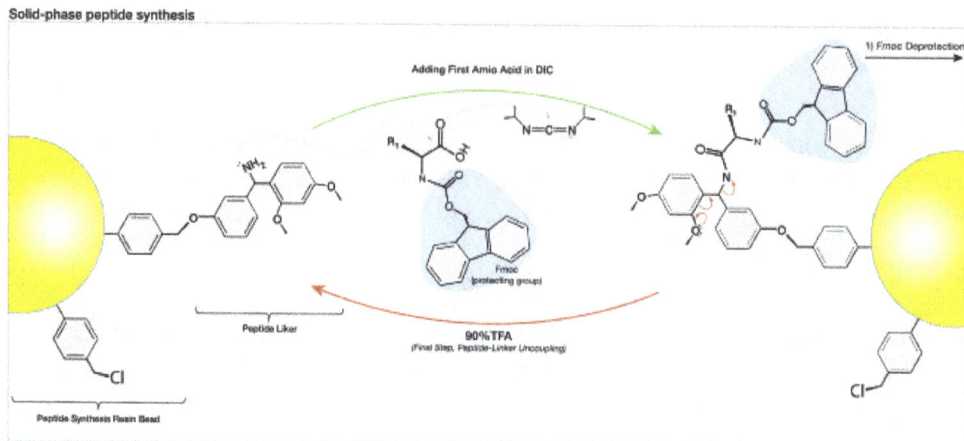

Solid-phase peptide synthesis on a rink amide resin using Fmoc-α-amine-protected amino acid

Peptides in Molecular Biology

Peptides received prominence in molecular biology for several reasons. The first is that peptides allow the creation of peptide antibodies in animals without the need of purifying the protein of interest. This involves synthesizing antigenic peptides of sections of the protein of interest. These will then be used to make antibodies in a rabbit or mouse against the protein.

Another reason is that peptides have become instrumental in mass spectrometry, allowing the identification of proteins of interest based on peptide masses and sequence.

Peptides have recently been used in the study of protein structure and function. For example, synthetic peptides can be used as probes to see where protein-peptide interactions occur.

Inhibitory peptides are also used in clinical research to examine the effects of peptides on the inhibition of cancer proteins and other diseases. For example, one of the most promising application is through peptides that target LHRH. These particular peptides act as an agonist, meaning that they bind to a cell in a way that regulates LHRH receptors. The process of inhibiting the cell receptors suggests that peptides could be beneficial in treating prostate cancer. However, additional investigations and experiments are required before the cancer-fighting attributes, exhibited by peptides, can be considered definitive.

Well-Known Peptide Families

The peptide families in this section are ribosomal peptides, usually with hormonal activity. All of these peptides are synthesized by cells as longer "propeptides" or "proproteins" and truncated prior to exiting the cell. They are released into the bloodstream where they perform their signaling functions.

Tachykinin Peptides

- Substance P
- Kassinin

- Neurokinin A
- Eledoisin
- Neurokinin B

Vasoactive Intestinal Peptides

- VIP (Vasoactive Intestinal Peptide; PHM27)
- PACAP Pituitary Adenylate Cyclase Activating Peptide
- Peptide PHI 27 (Peptide Histidine Isoleucine 27)
- GHRH 1-24 (Growth Hormone Releasing Hormone 1-24)
- Glucagon
- Secretin

Pancreatic Polypeptide-Related Peptides

- NPY (NeuroPeptide Y)
- PYY (Peptide YY)
- APP (Avian Pancreatic Polypeptide)
- PPY Pancreatic PolYpeptide

Opioid Peptides

- Proopiomelanocortin (POMC) peptides
- Enkephalin pentapeptides
- Prodynorphin peptides

Calcitonin Peptides

- Calcitonin
- Amylin
- AGG01

Other Peptides

- B-type Natriuretic Peptide (BNP) - produced in myocardium & useful in medical diagnosis
- Lactotripeptides - Lactotripeptides might reduce blood pressure, although the evidence is mixed.

Notes on Terminology

Length:

- A polypeptide is a single linear chain of many amino acids, held together by amide bonds.

- A protein is one or more polypeptide (more than about 50 amino acids long).

- An oligopeptide consists of only a few amino acids (between two and twenty).

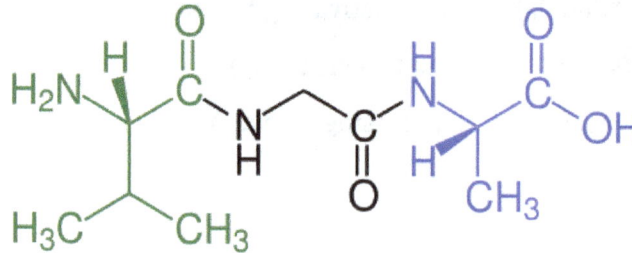

A tripeptide (example Val-Gly-Ala) with green marked amino end (L-Valine) and blue marked carboxyl end (L-Alanine)

Number of amino acids:

- A monopeptide has one amino acid.

- A dipeptide has two amino acids.

- A tripeptide has three amino acids.

- A tetrapeptide has four amino acids.

- A pentapeptide has five amino acids.

- A hexapeptide has six amino acids.

- A heptapeptide has seven amino acids.

- An octapeptide has eight amino acids (e.g., angiotensin II).

- A nonapeptide has nine amino acids (e.g., oxytocin).

- A decapeptide has ten amino acids (e.g., gonadotropin-releasing hormone & angiotensin I).

- An undecapeptide (or monodecapeptide) has eleven amino acids, a dodecapeptide (or didecapeptide) has twelve amino acids, a tridecapeptide has thirteen amino acids, and so forth.

- An icosapeptide has twenty amino acids, a tricontapeptide has thirty amino acids, a tetracontapeptide has forty amino acids, and so forth.

Function:

- A neuropeptide is a peptide that is active in association with neural tissue.

- A lipopeptide is a peptide that has a lipid connected to it, and pepducins are lipopeptides that interact with GPCRs.

- A peptide hormone is a peptide that acts as a hormone.

- A proteose is a mixture of peptides produced by the hydrolysis of proteins. The term is somewhat archaic.

Doping in Sports

The term peptide has been used to mean secretagogue peptides and peptide hormones in sports doping matters: secretagogue peptides are classified as Schedule 2 (S2) prohibited substances on the World Anti-Doping Agency (WADA) Prohibited List, and are therefore prohibited for use by professional athletes both in and out of competition. Such secretagogue peptides have been on the WADA prohibited substances list since at least 2008. The Australian Crime Commission cited the alleged misuse of secretagogue peptides in Australian sport including growth hormone releasing peptides CJC-1295, GHRP-6, and GHSR (gene) hexarelin. There is ongoing controversy on the legality of using secretagogue peptides in sports.

Generation of New Proteins

Protein Biosynthesis

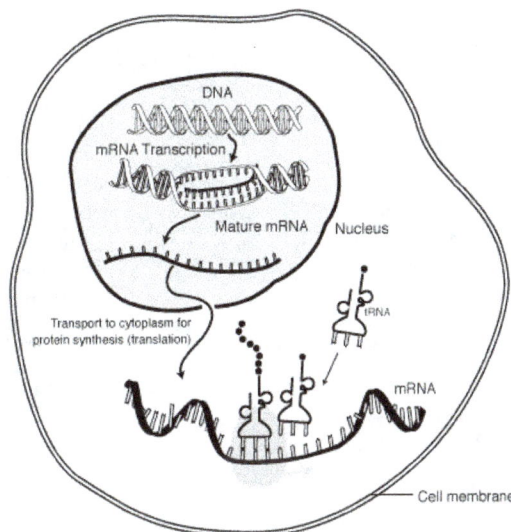

RNA is transcribed in the nucleus; once completely processed, it is transported to the cytoplasm and translated by the ribosome (shown in very pale grey behind the tRNA).

Protein biosynthesis is the process whereby biological cells generate new proteins; it is balanced by the loss of cellular proteins via degradation or export. Translation, the assembly of amino acids by ribosomes, is an essential part of the biosynthetic pathway, along with generation of messenger RNA (mRNA), aminoacylation of transfer RNA (tRNA), co-translational transport, and post-translational modification. Protein biosynthesis is strictly regulated at multiple steps. They are principally during transcription (phenomena of RNA synthesis from DNA template) and translation (phenomena of amino acid assembly from RNA).

The cistron DNA is transcribed into the first of a series of RNA intermediates. The last version is used as a template in synthesis of a polypeptide chain. Protein will often be synthesized directly from genes by translating mRNA. However, when a protein must be available on short notice or in large quantities, a protein precursor is produced. A proprotein is an inactive protein containing one or more inhibitory peptides that can be activated when the inhibitory sequence is removed by proteolysis during posttranslational modification. A preprotein is a form that contains a signal sequence (an N-terminal signal peptide) that specifies its insertion into or through membranes, i.e., targets them for secretion. The signal peptide is cleaved off in the endoplasmic reticulum. Pre-proproteins have both sequences (inhibitory and signal) still present.

In protein synthesis, a succession of tRNA molecules charged with appropriate amino acids are brought together with an mRNA molecule and matched up by base-pairing through the anti-codons of the tRNA with successive codons of the mRNA. The amino acids are then linked together to extend the growing protein chain, and the tRNAs, no longer carrying amino acids, are released. This whole complex of processes is carried out by the ribosome, formed of two main chains of RNA, called ribosomal RNA (rRNA), and more than 50 different proteins. The ribosome latches onto the end of an mRNA molecule and moves along it, capturing loaded tRNA molecules and joining together their amino acids to form a new protein chain.

Protein biosynthesis, although very similar, is different for prokaryotes and eukaryotes.

Transcription

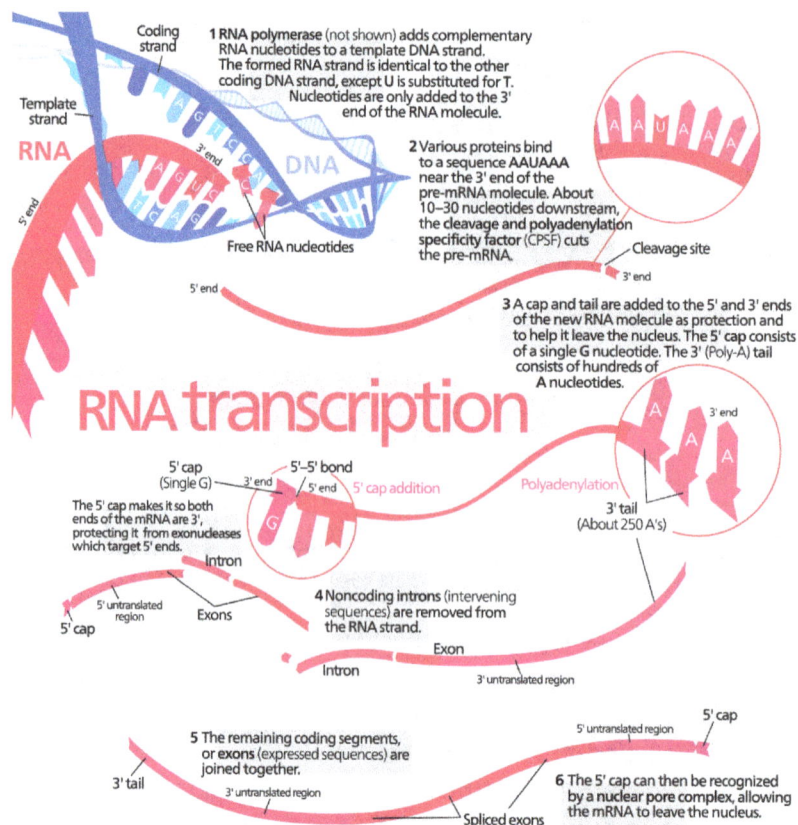

Diagram showing the process of transcription

In transcription an mRNA chain is generated, with one strand of the DNA double helix in the genome as a template. This strand is called the template strand. Transcription can be divided into 3 stages: initiation, elongation, and termination, each regulated by a large number of proteins such as transcription factors and coactivators that ensure that the correct gene is transcribed.

Transcription occurs in the cell nucleus, where the DNA is held. The DNA structure of the cell is made up of two helixes made up of sugar and phosphate held together by hydrogen bonds between the bases of opposite strands. The sugar and the phosphate in each strand are joined together by stronger phosphodiester covalent bonds. The DNA is "unzipped" (disruption of hydrogen bonds between different single strands) by the enzyme helicase, leaving the single nucleotide chain open to be copied. RNA polymerase reads the DNA strand from the 3-prime (3') end to the 5-prime (5') end, while it synthesizes a single strand of messenger RNA in the 5'-to-3' direction. The general RNA structure is very similar to the DNA structure, but in RNA the nucleotide uracil takes the place that thymine occupies in DNA. The single strand of mRNA leaves the nucleus through nuclear pores, and migrates into the cytoplasm.

The first product of transcription differs in prokaryotic cells from that of eukaryotic cells, as in prokaryotic cells the product is mRNA, which needs no post-transcriptional modification, whereas, in eukaryotic cells, the first product is called primary transcript, that needs post-transcriptional modification (capping with 7-methyl-guanosine, tailing with a poly A tail) to give hnRNA (heterogeneous nuclear RNA). hnRNA then undergoes splicing of introns (noncoding parts of the gene) via spliceosomes to produce the final mRNA.

Translation

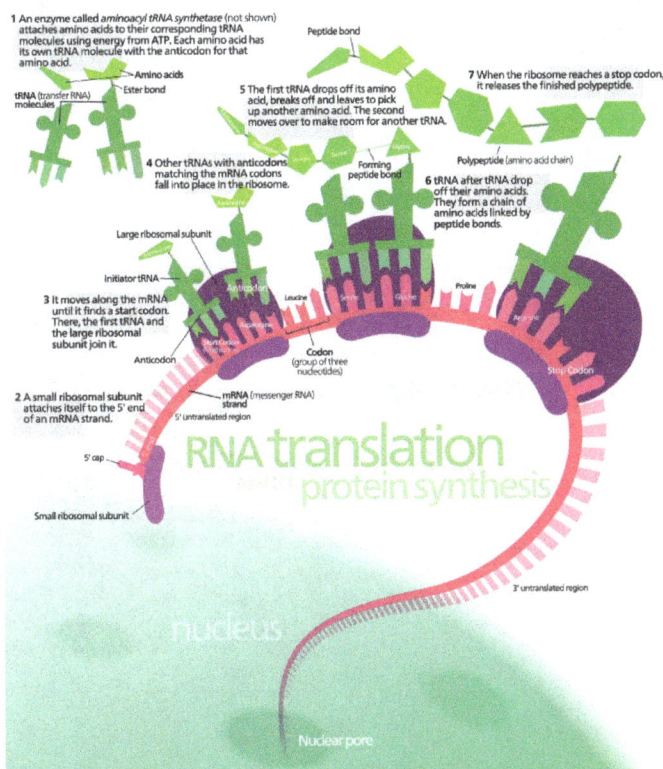

Diagram showing the process of translation

The synthesis of proteins from RNA is known as translation. In eukaryotes, translation occurs in the cytoplasm, where the ribosomes are located. Ribosomes are made of a small and large subunit that surround the mRNA. In translation, messenger RNA (mRNA) is decoded to produce a specific polypeptide according to the rules specified by the trinucleotide genetic code. This uses an mRNA sequence as a template to guide the synthesis of a chain of amino acids that form a protein. Translation proceeds in four phases: activation, initiation, elongation, and termination (all describing the growth of the amino acid chain, or polypeptide that is the product of translation).

In activation, the correct amino acid (AA) is joined to the correct transfer RNA (tRNA). While this is not, in the technical sense, a step in translation, it is required for translation to proceed. The AA is joined by its carboxyl group to the 3' OH of the tRNA by an ester bond. When the tRNA has an amino acid linked to it, it is termed "charged". Initiation involves the small subunit of the ribosome binding to 5' end of mRNA with the help of initiation factors (IF), other proteins that assist the process. Elongation occurs when the next aminoacyl-tRNA (charged tRNA) in line binds to the ribosome along with GTP and an elongation factor. Termination of the polypeptide happens when the A site of the ribosome faces a stop codon (UAA, UAG, or UGA). When this happens, no tRNA can recognize it, but releasing factor can recognize nonsense codons and causes the release of the polypeptide chain. The capacity of disabling or inhibiting translation in protein biosynthesis is used by some antibiotics such as anisomycin, cycloheximide, chloramphenicol, tetracycline, streptomycin, erythromycin, puromycin, etc.

Events Following Protein Translation

The events following biosynthesis include post-translational modification and protein folding. During and after synthesis, polypeptide chains often fold to assume, so called, native secondary and tertiary structures. This is known as protein folding.

Enzymes: An Integrated Study

This chapter discusses enzyme secretion and formation in living beings and also sheds light on the enzyme related activity. It provides detailed information about proteolysis and enzymes to the students. It discusses their structure, mechanism, etymology, inhibition and history. Students will get in-depth knowledge about enzymes from this chapter.

Enzyme

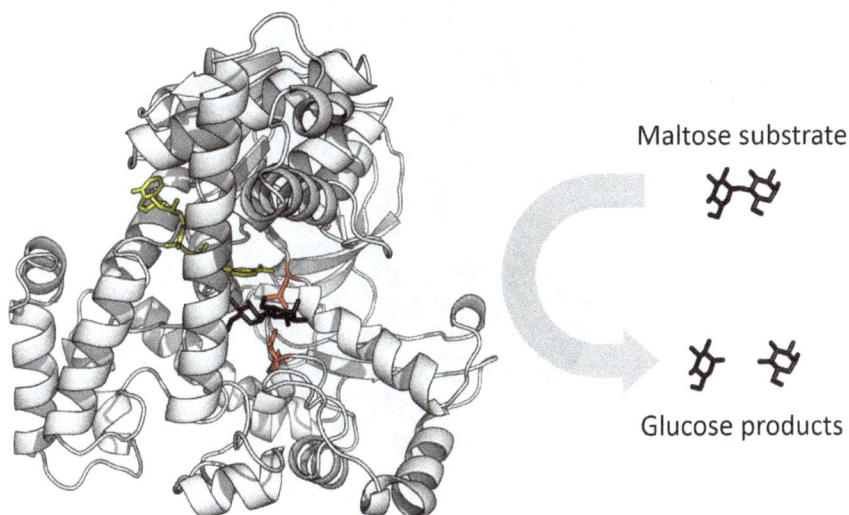

Maltose substrate

Glucose products

The enzyme glucosidase converts sugar maltose to two glucosesugars. Active site residues in red, maltose substrate in black, and NAD cofactor in yellow. (PDB: 1OBB)

Enzymes accelerate, or catalyze, chemical reactions. The molecules at the beginning of the process upon which enzymes may act are called substrates and the enzyme converts these into different molecules, called products. Almost all metabolic processes in the cell need enzymes in order to occur at rates fast enough to sustain life. The set of enzymes made in a cell determines which metabolic pathways occur in that cell. The study of enzymes is called enzymology.

Enzymes are known to catalyze more than 5,000 biochemical reaction types. Most enzymes are proteins, although a few are catalytic RNA molecules. Enzymes' specificity comes from their unique three-dimensional structures.

Like all catalysts, enzymes increase the rate of a reaction by lowering its activation energy. Some enzymes can make their conversion of substrate to product occur many millions of times faster. An extreme example is orotidine 5'-phosphate decarboxylase, which allows a reaction that would otherwise take millions of years to occur in milliseconds. Chemically, enzymes are like any catalyst and are not consumed in chemical reactions, nor do they alter the equilibrium of a reaction. Enzymes differ from most other catalysts by being much more specific. Enzyme activity can be affected by other molecules: inhibitors are molecules that decrease enzyme activity, and activators are molecules that increase activity. Many drugs and poisons are enzyme inhibitors. An enzyme's activity decreases markedly outside its optimal temperature and pH.

Some enzymes are used commercially, for example, in the synthesis of antibiotics. Some household products use enzymes to speed up chemical reactions: enzymes in biological washing powders break down protein, starch or fat stains on clothes, and enzymes in meat tenderizer break down proteins into smaller molecules, making the meat easier to chew.

Etymology and History

Eduard Buchner

By the late 17th and early 18th centuries, the digestion of meat by stomach secretions and the conversion of starch to sugars by plant extracts and saliva were known but the mechanisms by which these occurred had not been identified.

French chemist Anselme Payen was the first to discover an enzyme, diastase, in 1833. A few decades later, when studying the fermentation of sugar to alcohol by yeast, Louis Pasteur concluded that this fermentation was caused by a vital force contained within the yeast cells called "ferments", which were thought to function only within living organisms. He wrote that "alcoholic fermentation is an act correlated with the life and organization of the yeast cells, not with the death or putrefaction of the cells."

In 1877, German physiologist Wilhelm Kühne (1837–1900) first used the term enzyme, to describe this process. The word enzyme was used later to refer to nonliving substances such as pepsin, and the word ferment was used to refer to chemical activity produced by living organisms.

Eduard Buchner submitted his first paper on the study of yeast extracts in 1897. In a series of experiments at the University of Berlin, he found that sugar was fermented by yeast extracts even when there were no living yeast cells in the mixture. He named the enzyme that brought about the fermentation of sucrose "zymase". In 1907, he received the Nobel Prize in Chemistry for "his discovery of cell-free fermentation". Following Buchner's example, enzymes are usually named according to the reaction they carry out: the suffix -ase is combined with the name of the substrate (e.g., lactase is the enzyme that cleaves lactose) or to the type of reaction (e.g., DNA polymerase forms DNA polymers).

The biochemical identity of enzymes was still unknown in the early 1900s. Many scientists observed that enzymatic activity was associated with proteins, but others (such as Nobel laureate Richard Willstätter) argued that proteins were merely carriers for the true enzymes and that proteins per se were incapable of catalysis. In 1926, James B. Sumner showed that the enzyme urease was a pure protein and crystallized it; he did likewise for the enzyme catalase in 1937. The conclusion that pure proteins can be enzymes was definitively demonstrated by John Howard Northrop and Wendell Meredith Stanley, who worked on the digestive enzymes pepsin (1930), trypsin and chymotrypsin. These three scientists were awarded the 1946 Nobel Prize in Chemistry.

The discovery that enzymes could be crystallized eventually allowed their structures to be solved by x-ray crystallography. This was first done for lysozyme, an enzyme found in tears, saliva and egg whites that digests the coating of some bacteria; the structure was solved by a group led by David Chilton Phillips and published in 1965. This high-resolution structure of lysozyme marked the beginning of the field of structural biology and the effort to understand how enzymes work at an atomic level of detail.

Structure

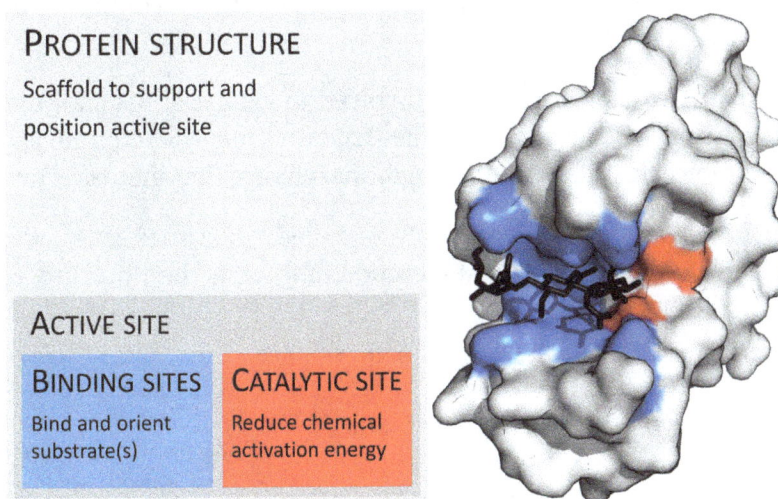

Organisation of enzyme structure and lysozyme example. Binding sites in blue, catalytic site in red and peptidoglycan substrate in black. (PDB: 9LYZ)

Enzymes are generally globular proteins, acting alone or in larger complexes. Like all proteins, enzymes are linear chains of amino acids that fold to produce a three-dimensional structure. The sequence of the amino acids specifies the structure which in turn determines the catalytic activity of the enzyme. Although structure determines function, a novel enzyme's activity cannot yet be predicted from its structure alone. Enzyme structures unfold (denature) when heated or exposed to chemical denaturants and this disruption to the structure typically causes a loss of activity. Enzyme denaturation is normally linked to temperatures above a species' normal level; as a result, enzymes from bacteria living in volcanic environments such as hot springs are prized by industrial users for their ability to function at high temperatures, allowing enzyme-catalysed reactions to be operated at a very high rate.

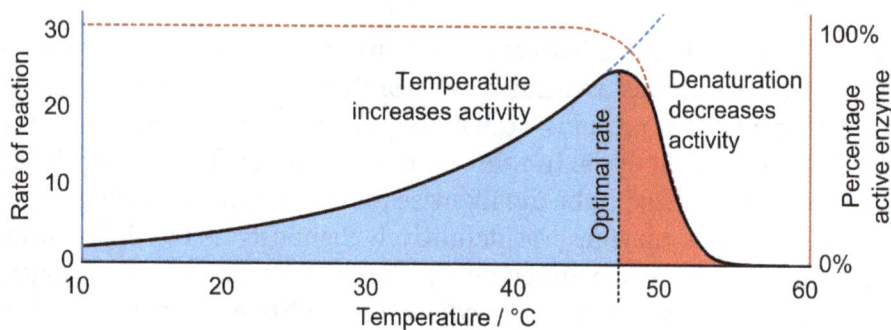

Enzyme activity initially increases with temperature (Q10 coefficient) until the enzyme's structure unfolds (denaturation), leading to an optimal rate of reaction at an intermediate temperature.

Enzymes are usually much larger than their substrates. Sizes range from just 62 amino acid residues, for the monomer of 4-oxalocrotonate tautomerase, to over 2,500 residues in the animal fatty acid synthase. Only a small portion of their structure (around 2–4 amino acids) is directly involved in catalysis: the catalytic site. This catalytic site is located next to one or more binding sites where residues orient the substrates. The catalytic site and binding site together comprise the enzyme's active site. The remaining majority of the enzyme structure serves to maintain the precise orientation and dynamics of the active site.

In some enzymes, no amino acids are directly involved in catalysis; instead, the enzyme contains sites to bind and orient catalytic cofactors. Enzyme structures may also contain allosteric sites where the binding of a small molecule causes a conformational change that increases or decreases activity.

A small number of RNA-based biological catalysts called ribozymes exist, which again can act alone or in complex with proteins. The most common of these is the ribosome which is a complex of protein and catalytic RNA components.

Mechanism

Substrate Binding

Enzymes must bind their substrates before they can catalyse any chemical reaction. Enzymes are usually very specific as to what substrates they bind and then the chemical reaction catalysed. Specificity is achieved by binding pockets with complementary shape, charge and hydrophilic/hydrophobic characteristics to the substrates. Enzymes can therefore distinguish between very similar substrate molecules to be chemoselective, regioselective and stereospecific.

Some of the enzymes showing the highest specificity and accuracy are involved in the copying and expression of the genome. Some of these enzymes have "proof-reading" mechanisms. Here, an enzyme such as DNA polymerase catalyzes a reaction in a first step and then checks that the product is correct in a second step. This two-step process results in average error rates of less than 1 error in 100 million reactions in high-fidelity mammalian polymerases.·· Similar proofreading mechanisms are also found in RNA polymerase, aminoacyl tRNA synthetases and ribosomes.

Conversely, some enzymes display enzyme promiscuity, having broad specificity and acting on a range of different physiologically relevant substrates. Many enzymes possess small side activities which arose fortuitously (i.e. neutrally), which may be the starting point for the evolutionary selection of a new function.

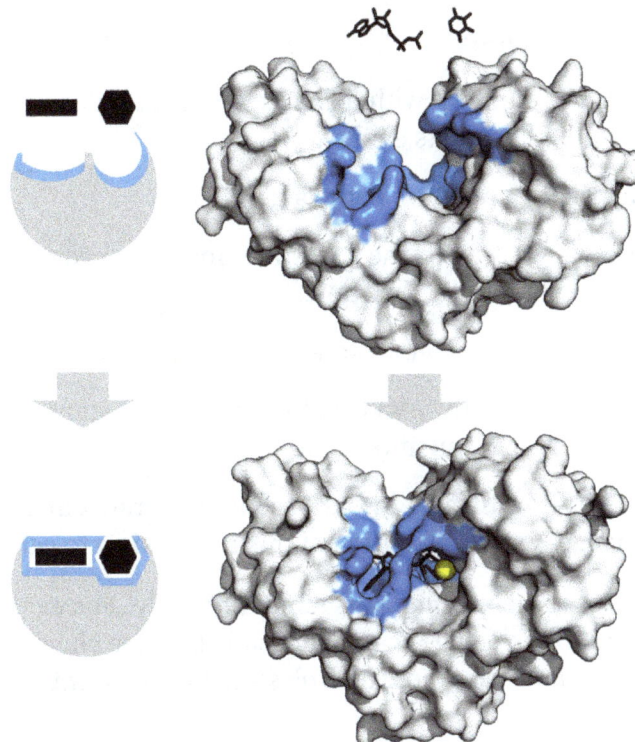

Enzyme changes shape by induced fit upon substrate binding to form enzyme-substrate complex. Hexokinase has a large induced fit motion that closes over the substrates adenosine triphosphate and xylose. Binding sites in blue, substrates in black and Mg+ cofactor in yellow. (PDB: 2E2N, 2E2Q)

"Lock and Key" Model

To explain the observed specificity of enzymes, in 1894 Emil Fischer proposed that both the enzyme and the substrate possess specific complementary geometric shapes that fit exactly into one another. This is often referred to as "the lock and key" model.·· This early model explains enzyme specificity, but fails to explain the stabilization of the transition state that enzymes achieve.

Induced Fit Model

In 1958, Daniel Koshland suggested a modification to the lock and key model: since enzymes are rather flexible structures, the active site is continuously reshaped by interactions with the substrate as the substrate interacts with the enzyme. As a result, the substrate does not simply bind to a rigid active site; the amino acid side-chains that make up the active site are molded into the precise positions that enable the enzyme to perform its catalytic function. In some cases, such as glycosidases, the substrate molecule also changes shape slightly as it enters the active site. The active site continues to change until the substrate is completely bound, at which point the final shape and charge distribution is determined. Induced fit may enhance the fidelity of molecular recognition in the presence of competition and noise via the conformational proofreading mechanism.

Catalysis

Enzymes can accelerate reactions in several ways, all of which lower the activation energy ($\Delta G^{\ddagger}$, Gibbs free energy)

1. By stabilizing the transition state:

    o   Creating an environment with a charge distribution complementary to that of the transition state to lower its energy.

2. By providing an alternative reaction pathway:

    o   Temporarily reacting with the substrate, forming a covalent intermediate to provide a lower energy transition state.

3. By destabilising the substrate ground state:

    o   Distorting bound substrate(s) into their transition state form to reduce the energy required to reach the transition state.

    o   By orienting the substrates into a productive arrangement to reduce the reaction entropy change. The contribution of this mechanism to catalysis is relatively small.

Enzymes may use several of these mechanisms simultaneously. For example, proteases such as trypsin perform covalent catalysis using a catalytic triad, stabilise charge build-up on the transition states using an oxyanion hole, complete hydrolysis using an oriented water substrate.

Dynamics

Enzymes are not rigid, static structures; instead they have complex internal dynamic motions – that is, movements of parts of the enzyme's structure such as individual amino acid residues, groups of residues forming a protein loop or unit of secondary structure, or even an entire protein domain. These motions give rise to a conformational ensemble of slightly different structures that interconvert with one another at equilibrium. Different states within this ensemble may be associated with different aspects of an enzyme's function. For example, different conformations of the enzyme dihydrofolate reductase are associated with the substrate binding, catalysis, cofactor release, and product release steps of the catalytic cycle.

Allosteric Modulation

Allosteric sites are pockets on the enzyme, distinct from the active site, that bind to molecules in the cellular environment. These molecules then cause a change in the conformation or dynamics of the enzyme that is transduced to the active site and thus affects the reaction rate of the enzyme. In this way, allosteric interactions can either inhibit or activate enzymes. Allosteric interactions with metabolites upstream or downstream in an enzyme's metabolic pathway cause feedback regulation, altering the activity of the enzyme according to the flux through the rest of the pathway.

Cofactors

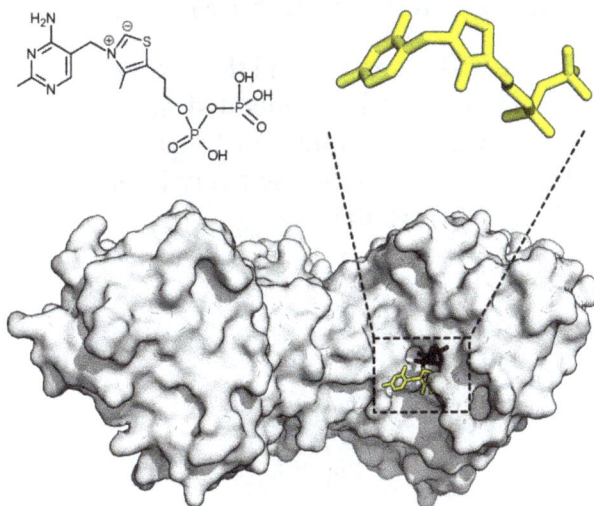

Chemical structure for thiamine pyrophosphate and protein structure of transketolase. Thiamine pyrophosphate cofactor in yellow and xylulose 5-phosphate substrate in black. (PDB: 4KXV)

Some enzymes do not need additional components to show full activity. Others require non-protein molecules called cofactors to be bound for activity. Cofactors can be either inorganic (e.g., metal ions and iron-sulfur clusters) or organic compounds (e.g., flavin and heme). Organic cofactors can be either coenzymes, which are released from the enzyme's active site during the reaction, or prosthetic groups, which are tightly bound to an enzyme. Organic prosthetic groups can be covalently bound (e.g., biotin in enzymes such as pyruvate carboxylase).

An example of an enzyme that contains a cofactor is carbonic anhydrase, which is shown in the ribbon diagram above with a zinc cofactor bound as part of its active site. These tightly bound ions or molecules are usually found in the active site and are involved in catalysis.For example, flavin and heme cofactors are often involved in redox reactions.

Enzymes that require a cofactor but do not have one bound are called apoenzymes or apoproteins. An enzyme together with the cofactor(s) required for activity is called a holoenzyme (or haloenzyme). The term holoenzyme can also be applied to enzymes that contain multiple protein subunits, such as the DNA polymerases; here the holoenzyme is the complete complex containing all the subunits needed for activity.

Coenzymes

Coenzymes are small organic molecules that can be loosely or tightly bound to an enzyme. Coenzymes transport chemical groups from one enzyme to another. Examples include NADH, NADPH and adenosine triphosphate (ATP). Some coenzymes, such as riboflavin, thiamine and folic acid, are vitamins, or compounds that cannot be synthesized by the body and must be acquired from the diet. The chemical groups carried include the hydride ion (H^-) carried by NAD or $NADP^+$, the phosphate group carried by adenosine triphosphate, the acetyl group carried by coenzyme A, formyl, methenyl or methyl groups carried by folic acid and the methyl group carried by S-adenosylmethionine.

Since coenzymes are chemically changed as a consequence of enzyme action, it is useful to consider coenzymes to be a special class of substrates, or second substrates, which are common to many different enzymes. For example, about 1000 enzymes are known to use the coenzyme NADH.

Coenzymes are usually continuously regenerated and their concentrations maintained at a steady level inside the cell. For example, NADPH is regenerated through the pentose phosphate pathway and S-adenosylmethionine by methionine adenosyltransferase. This continuous regeneration means that small amounts of coenzymes can be used very intensively. For example, the human body turns over its own weight in ATP each day.[

Kinetics

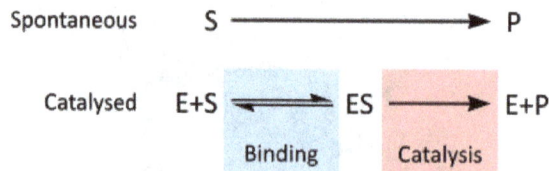

A chemical reaction mechanism with or without enzyme catalysis. The enzyme (E) binds substrate (S) to produce product (P).

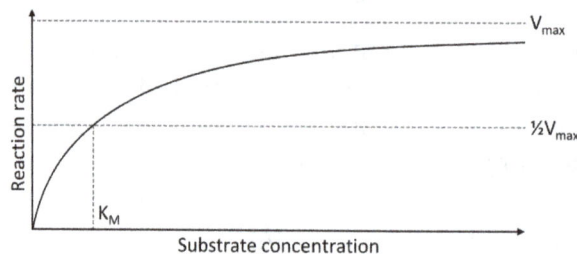

Saturation curve for an enzyme reaction showing the relation between the substrate concentration and reaction rate.

Enzyme kinetics is the investigation of how enzymes bind substrates and turn them into products. The rate data used in kinetic analyses are commonly obtained from enzyme assays. In 1913 Leonor Michaelis and Maud Leonora Menten proposed a quantitative theory of enzyme kinetics, which is referred to as Michaelis–Menten kinetics. The major contribution of Michaelis and Menten was to think of enzyme reactions in two stages. In the first, the substrate binds reversibly to the enzyme, forming the enzyme-substrate complex. This is sometimes called the Michaelis-Menten complex in their honor. The enzyme then catalyzes the chemical step in the reaction and releases the product. This work was further developed by G. E. Briggs and J. B. S. Haldane, who derived kinetic equations that are still widely used today.

Enzyme rates depend on solution conditions and substrate concentration. To find the maximum speed of an enzymatic reaction, the substrate concentration is increased until a constant rate of product formation is seen. This is shown in the saturation curve on the right. Saturation happens because, as substrate concentration increases, more and more of the free enzyme is converted into the substrate-bound ES complex. At the maximum reaction rate (V_{max}) of the enzyme, all the enzyme active sites are bound to substrate, and the amount of ES complex is the same as the total amount of enzyme.·

V_{max} is only one of several important kinetic parameters. The amount of substrate needed to achieve a given rate of reaction is also important. This is given by the Michaelis-Menten constant

(K_m), which is the substrate concentration required for an enzyme to reach one-half its maximum reaction rate; generally, each enzyme has a characteristic K_m for a given substrate. Another useful constant is k_{cat}, also called the turnover number, which is the number of substrate molecules handled by one active site per second.·

The efficiency of an enzyme can be expressed in terms of k_{cat}/K_m. This is also called the specificity constant and incorporates the rate constants for all steps in the reaction up to and including the first irreversible step. Because the specificity constant reflects both affinity and catalytic ability, it is useful for comparing different enzymes against each other, or the same enzyme with different substrates. The theoretical maximum for the specificity constant is called the diffusion limit and is about 10 to 10 (M⁻ s⁻). At this point every collision of the enzyme with its substrate will result in catalysis, and the rate of product formation is not limited by the reaction rate but by the diffusion rate. Enzymes with this property are called catalytically perfect or kinetically perfect. Example of such enzymes are triose-phosphate isomerase, carbonic anhydrase, acetylcholinesterase, catalase, fumarase, β-lactamase, and superoxide dismutase.·· The turnover of such enzymes can reach several million reactions per second.·

Michaelis–Menten kinetics relies on the law of mass action, which is derived from the assumptions of free diffusion and thermodynamically driven random collision. Many biochemical or cellular processes deviate significantly from these conditions, because of macromolecular crowding and constrained molecular movement. More recent, complex extensions of the model attempt to correct for these effects.

Inhibition

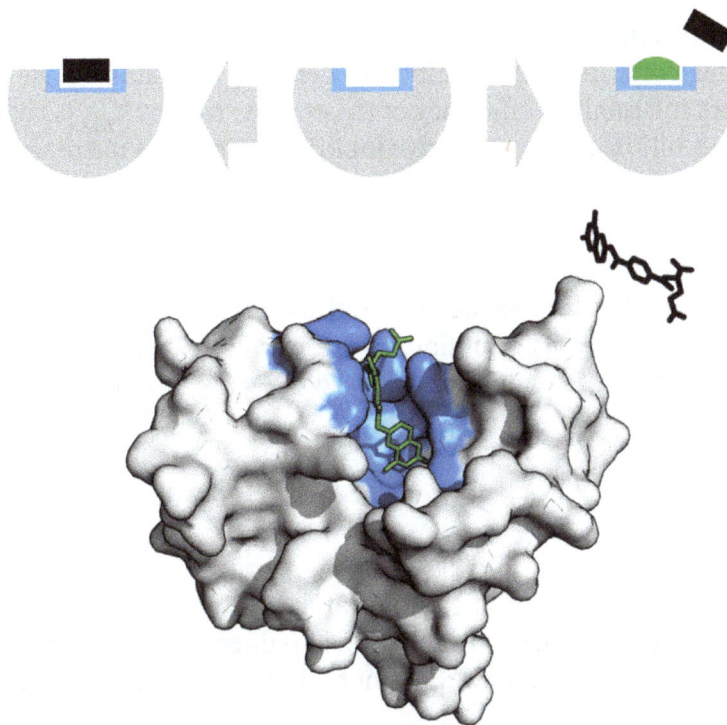

An enzyme binding site that would normally bind substrate can alternatively bind a competitive inhibitor, preventing substrate access. Dihydrofolate reductase is inhibited by methotrexate which prevents binding of its substrate, folic acid. Binding site in blue, inhibitor in green, and substrate in black. (PDB: 4QI9)

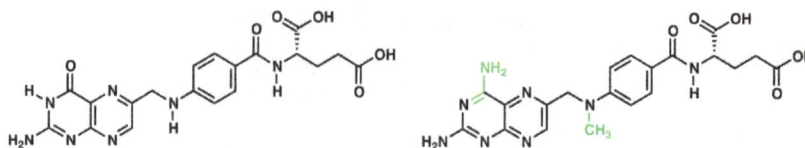

The coenzyme folic acid (left) and the anti-cancer drug methotrexate (right) are very similar in structure (differences show in green). As a result, methotrexate is a competitive inhibitor of many enzymes that use folates.

Enzyme reaction rates can be decreased by various types of enzyme inhibitors.

Types of Inhibition

Competitive

A competitive inhibitor and substrate cannot bind to the enzyme at the same time. Often competitive inhibitors strongly resemble the real substrate of the enzyme. For example, the drug methotrexate is a competitive inhibitor of the enzyme dihydrofolate reductase, which catalyzes the reduction of dihydrofolate to tetrahydrofolate. The similarity between the structures of dihydrofolate and this drug are shown in the accompanying figure. This type of inhibition can be overcome with high substrate concentration. In some cases, the inhibitor can bind to a site other than the binding-site of the usual substrate and exert an allosteric effect to change the shape of the usual binding-site.

Non-Competitive

A non-competitive inhibitor binds to a site other than where the substrate binds. The substrate still binds with its usual affinity and hence K_m remains the same. However the inhibitor reduces the catalytic efficiency of the enzyme so that V_{max} is reduced. In contrast to competitive inhibition, non-competitive inhibition cannot be overcome with high substrate concentration.¯

Uncompetitive

An uncompetitive inhibitor cannot bind to the free enzyme, only to the enzyme-substrate complex; hence, these types of inhibitors are most effective at high substrate concentration. In the presence of the inhibitor, the enzyme-substrate complex is inactive. This type of inhibition is rare.

Mixed

A mixed inhibitor binds to an allosteric site and the binding of the substrate and the inhibitor affect each other. The enzyme's function is reduced but not eliminated when bound to the inhibitor. This type of inhibitor does not follow the Michaelis-Menten equation.

Irreversible

An irreversible inhibitor permanently inactivates the enzyme, usually by forming a covalent bond to the protein. Penicillin and aspirin are common drugs that act in this manner.

Functions of Inhibitors

In many organisms, inhibitors may act as part of a feedback mechanism. If an enzyme produces too much of one substance in the organism, that substance may act as an inhibitor for the enzyme at the beginning of the pathway that produces it, causing production of the substance to slow down or stop when there is sufficient amount. This is a form of negative feedback. Major metabolic pathways such as the citric acid cycle make use of this mechanism.

Since inhibitors modulate the function of enzymes they are often used as drugs. Many such drugs are reversible competitive inhibitors that resemble the enzyme's native substrate, similar to methotrexate above; other well-known examples include statins used to treat high cholesterol, and protease inhibitors used to treat retroviral infections such as HIV. A common example of an irreversible inhibitor that is used as a drug is aspirin, which inhibits the COX-1 and COX-2 enzymes that produce the inflammation messenger prostaglandin. Other enzyme inhibitors are poisons. For example, the poison cyanide is an irreversible enzyme inhibitor that combines with the copper and iron in the active site of the enzyme cytochrome c oxidase and blocks cellular respiration.

Biological Function

Enzymes serve a wide variety of functions inside living organisms. They are indispensable for signal transduction and cell regulation, often via kinases and phosphatases. They also generate movement, with myosin hydrolyzing ATP to generate muscle contraction, and also transport cargo around the cell as part of the cytoskeleton. Other ATPases in the cell membrane are ion pumps involved in active transport. Enzymes are also involved in more exotic functions, such as luciferase generating light in fireflies. Viruses can also contain enzymes for infecting cells, such as the HIV integrase and reverse transcriptase, or for viral release from cells, like the influenza virus neuraminidase.

An important function of enzymes is in the digestive systems of animals. Enzymes such as amylases and proteases break down large molecules (starch or proteins, respectively) into smaller ones, so they can be absorbed by the intestines. Starch molecules, for example, are too large to be absorbed from the intestine, but enzymes hydrolyze the starch chains into smaller molecules such as maltose and eventually glucose, which can then be absorbed. Different enzymes digest different food substances. In ruminants, which have herbivorous diets, microorganisms in the gut produce another enzyme, cellulase, to break down the cellulose cell walls of plant fiber.

Metabolism

Several enzymes can work together in a specific order, creating metabolic pathways. In a metabolic pathway, one enzyme takes the product of another enzyme as a substrate. After the catalytic reaction, the product is then passed on to another enzyme. Sometimes more than one enzyme can catalyze the same reaction in parallel; this can allow more complex regulation: with, for example, a low constant activity provided by one enzyme but an inducible high activity from a second enzyme.

Enzymes determine what steps occur in these pathways. Without enzymes, metabolism would neither progress through the same steps and could not be regulated to serve the needs of the cell. Most central metabolic pathways are regulated at a few key steps, typically through enzymes whose activity involves the hydrolysis of ATP. Because this reaction releases so much energy, other reactions that are thermodynamically unfavorable can be coupled to ATP hydrolysis, driving the overall series of linked metabolic reactions.

Control of Activity

There are five main ways that enzyme activity is controlled in the cell.

Regulation

Enzymes can be either activated or inhibited by other molecules. For example, the end product(s) of a metabolic pathway are often inhibitors for one of the first enzymes of the pathway (usually the first irreversible step, called committed step), thus regulating the amount of end product made by the pathways. Such a regulatory mechanism is called a negative feedback mechanism, because the amount of the end product produced is regulated by its own concentration.⁻ Negative feedback mechanism can effectively adjust the rate of synthesis of intermediate metabolites according to the demands of the cells. This helps with effective allocations of materials and energy economy, and it prevents the excess manufacture of end products. Like other homeostatic devices, the control of enzymatic action helps to maintain a stable internal environment in living organisms.

Post-Translational Modification

Examples of post-translational modification include phosphorylation, myristoylation and glycosylation.⁻ For example, in the response to insulin, the phosphorylation of multiple enzymes, including glycogen synthase, helps control the synthesis or degradation of glycogen and allows the cell to respond to changes in blood sugar. Another example of post-translational modification is the cleavage of the polypeptide chain. Chymotrypsin, a digestive protease, is produced in inactive form as chymotrypsinogen in the pancreas and transported in this form to the stomach where it is activated. This stops the enzyme from digesting the pancreas or other tissues before it enters the gut. This type of inactive precursor to an enzyme is known as a zymogen⁻ or proenzyme.

Quantity

Enzyme production (transcription and translation of enzyme genes) can be enhanced or diminished by a cell in response to changes in the cell's environment. This form of gene regulation is called enzyme induction. For example, bacteria may become resistant to antibiotics such as penicillin because enzymes called beta-lactamases are induced that hydrolyse the crucial beta-lactam ring within the penicillin molecule. Another example comes from enzymes in the liver called cytochrome P450 oxidases, which are important in drug metabolism. Induction or inhibition of these enzymes can cause drug interactions. Enzyme levels can also be regulated by changing the rate of enzyme degradation.

Subcellular Distribution

Enzymes can be compartmentalized, with different metabolic pathways occurring in different cellular compartments. For example, fatty acids are synthesized by one set of enzymes in the cytosol, endoplasmic reticulum and Golgi and used by a different set of enzymes as a source of energy in the mitochondrion, through β-oxidation. In addition, trafficking of the enzyme to different compartments may change the degree of protonation (cytoplasm neutral and lysosome acidic) or oxidative state [e.g., oxidized (periplasm) or reduced (cytoplasm)] which in turn affects enzyme activity.

Organ Specialization

In multicellular eukaryotes, cells in different organs and tissues have different patterns of gene expression and therefore have different sets of enzymes (known as isozymes) available for metabolic reactions. This provides a mechanism for regulating the overall metabolism of the organism. For example, hexokinase, the first enzyme in the glycolysis pathway, has a specialized form called glucokinase expressed in the liver and pancreas that has a lower affinity for glucose yet is more sensitive to glucose concentration. This enzyme is involved in sensing blood sugar and regulating insulin production.

Involvement in Disease

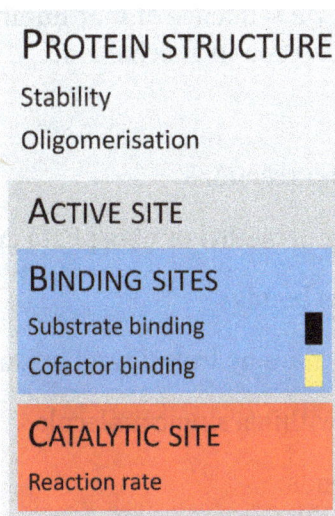

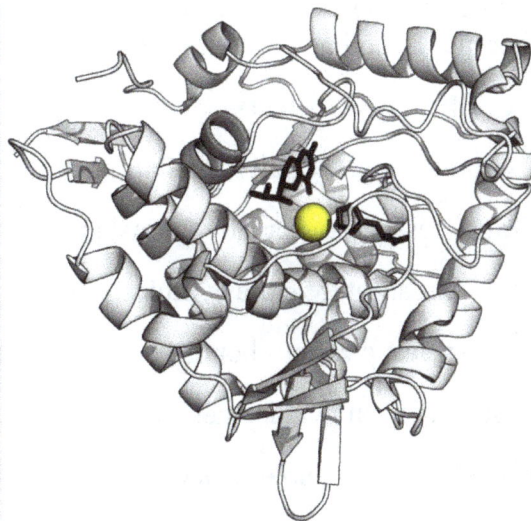

In phenylalanine hydroxylase over 300 different mutations throughout the structure cause phenylketonuria. Phenylalanine substrate and tetrahydrobiopterin coenzyme in black, and Fe⁺ cofactor in yellow. (PDB: 1KW0)

Since the tight control of enzyme activity is essential for homeostasis, any malfunction (mutation, overproduction, underproduction or deletion) of a single critical enzyme can lead to a genetic disease. The malfunction of just one type of enzyme out of the thousands of types present in the human body can be fatal. An example of a fatal genetic disease due to enzyme insufficiency is Tay-Sachs disease, in which patients lack the enzyme hexosaminidase.

One example of enzyme deficiency is the most common type of phenylketonuria. Many different single amino acid mutations in the enzyme phenylalanine hydroxylase, which catalyzes the first step in the degradation of phenylalanine, result in build-up of phenylalanine and related products. Some mutations are in the active site, directly disrupting binding and catalysis, but many are far from the active site and reduce activity by destabilising the protein structure, or affecting correct oligomerisation. This can lead to intellectual disability if the disease is untreated. Another example is pseudocholinesterase deficiency, in which the body's ability to break down choline ester drugs is impaired. Oral administration of enzymes can be used to treat some functional enzyme deficiencies, such as pancreatic insufficiency and lactose intolerance.

Another way enzyme malfunctions can cause disease comes from germline mutations in genes coding for DNA repair enzymes. Defects in these enzymes cause cancer because cells are less able to repair mutations in their genomes. This causes a slow accumulation of mutations and results in the development of cancers. An example of such a hereditary cancer syndrome is xeroderma pigmentosum, which causes the development of skin cancers in response to even minimal exposure to ultraviolet light.

Naming Conventions

An enzyme's name is often derived from its substrate or the chemical reaction it catalyzes, with the word ending in -ase. Examples are lactase, alcohol dehydrogenase and DNA polymerase. Different enzymes that catalyze the same chemical reaction are called isozymes.

The International Union of Biochemistry and Molecular Biology have developed a nomenclature for enzymes, the EC numbers; each enzyme is described by a sequence of four numbers preceded by "EC". The first number broadly classifies the enzyme based on its mechanism.

The top-Level Classification is:

- EC 1, Oxidoreductases: catalyze oxidation/reduction reactions
- EC 2, Transferases: transfer a functional group (e.g. a methyl or phosphate group)
- EC 3, Hydrolases: catalyze the hydrolysis of various bonds
- EC 4, Lyases: cleave various bonds by means other than hydrolysis and oxidation
- EC 5, Isomerases: catalyze isomerization changes within a single molecule
- EC 6, Ligases: join two molecules with covalent bonds.

These sections are subdivided by other features such as the substrate, products, and chemical mechanism. An enzyme is fully specified by four numerical designations. For example, hexokinase (EC 2.7.1.1) is a transferase (EC 2) that adds a phosphate group (EC 2.7) to a hexose sugar, a molecule containing an alcohol group (EC 2.7.1).

Industrial Applications

Enzymes are used in the chemical industry and other industrial applications when extremely specific catalysts are required. Enzymes in general are limited in the number of reactions they have evolved to catalyze and also by their lack of stability in organic solvents and at high temperatures. As a consequence, protein engineering is an active area of research and involves attempts to create new enzymes with novel properties, either through rational design or in vitro evolution. These efforts have begun to be successful, and a few enzymes have now been designed "from scratch" to catalyze reactions that do not occur in nature.

Application	Enzymes used	Uses
Biofuel industry	Cellulases	Break down cellulose into sugars that can be fermented to produce cellulosic ethanol.
	Ligninases	Pretreatment of biomass for biofuel production.
Biological detergent	Proteases, amylases, lipases	Remove protein, starch, and fat or oil stains from laundry and dishware.
	Mannanases	Remove food stains from the common food additive guar gum.
Brewing industry	Amylase, glucanases, proteases	Split polysaccharides and proteins in the malt.
	Betaglucanases	Improve the wort and beer filtration characteristics.
	Amyloglucosidase and pullulanases	Make low-calorie beer and adjust fermentability.
	Acetolactate decarboxylase (ALDC)	Increase fermentation efficiency by reducing diacetyl formation.
Culinary uses	Papain	Tenderize meat for cooking.
Dairy industry	Rennin	Hydrolyze protein in the manufacture of cheese.
	Lipases	Produce Camembert cheese and blue cheeses such as Roquefort.
Food processing	Amylases	Produce sugars from starch, such as in making high-fructose corn syrup.
	Proteases	Lower the protein level of flour, as in biscuit-making.
	Trypsin	Manufacture hypoallergenic baby foods.
	Cellulases, pectinases	Clarify fruit juices.
Molecular biology	Nucleases, DNA ligase and polymerases	Use restriction digestion and the polymerase chain reaction to create recombinant DNA.
Paper industry	Xylanases, hemicellulases and lignin peroxidases	Remove lignin from kraft pulp.
Personal care	Proteases	Remove proteins on contact lenses to prevent infections.
Starch industry	Amylases	Convert starch into glucose and various syrups.

Proteolysis

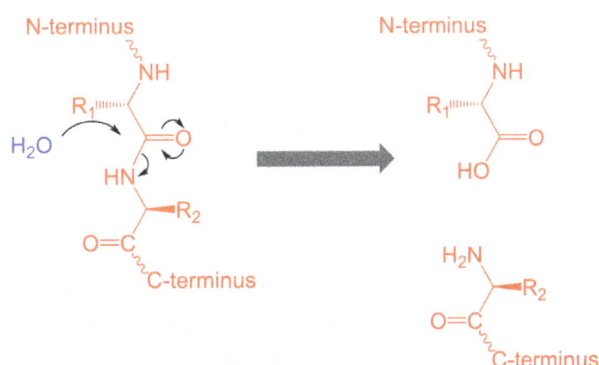

The hydrolysis of a protein (red) by the nucleophilic attack of water (blue). The uncatalysed half-life is several hundred years.

Proteolysis is the breakdown of proteins into smaller polypeptides or amino acids. Uncatalysed, the hydrolysis of peptide bonds is extremely slow, taking hundreds of years. Proteolysis is typically catalysed by cellular enzymes called proteases, but may also occur by intra-molecular digestion. Low pH or high temperatures can also cause proteolysis non-enzymatically.

Proteolysis in organisms serves many purposes; for example, digestive enzymes break down proteins in food to provide amino acids for the organism, while proteolytic processing of a polypeptide chain after its synthesis may be necessary for the production of an active protein. It is also important in the regulation of some physiological and cellular processes, as well as preventing the accumulation of unwanted or abnormal proteins in cells. Consequently, dis-regulation of proteolysis can cause diseases and is used in some venoms to damage their prey.

Proteolysis is important as an analytical tool for studying proteins in the laboratory, as well as industrially, for example in food processing and stain removal.

Biological Functions

Post-Translational Proteolytic Processing

Limited proteolysis of a polypeptide during or after translation in protein synthesis often occurs for many proteins. This may involve removal of the N-terminal methionine, signal peptide, and/or the conversion of an inactive or non-functional protein to an active one. The precursor to the final functional form of protein is termed proprotein, and these proproteins may be first synthesized as preproprotein. For example, albumin is first synthesized as preproalbumin and contains an uncleaved signal peptide. This forms the proalbumin after the signal peptide is cleaved, and a further processing to remove the N-terminal 6-residue propeptide yields the mature form of the protein.

Removal of N-Terminal Methionine

The initiating methionine (and, in prokaryotes, fMet) may be removed during translation of the nascent protein. For E. coli, fMet is efficiently removed if the second residue is small and uncharged, but not if the second residue is bulky and charged. In both prokaryotes and eukaryotes, the exposed N-terminal residue may determine the half-life of the protein according to the N-end rule.

Removal of the Signal Sequence

Proteins that are to be targeted to a particular organelle or for secretion have an N-terminal signal peptide that directs the protein to its final destination. This signal peptide is removed by proteolysis after their transport through a membrane.

Cleavage of Polyproteins

Some proteins and most eukaryotic polypeptide hormones are synthesized as a large precursor polypeptide known as a polyprotein that requires proteolytic cleavage into individual smaller polypeptide chains. The polyprotein pro-opiomelanocortin (POMC) contains many polypeptide hormones. The cleavage pattern of POMC, however, may vary between different tissues, yielding different sets of polypeptide hormones from the same polyprotein.

Many viruses also produce their proteins initially as a single polypeptide chain that were translated from a polycistronic mRNA. This polypeptide is subsequently cleaved into individual polypeptide chains.

Cleavage of Precursor Proteins

Many proteins and hormones are synthesized in the form of their precursors - zymogens, proenzymes, and prehormones. These proteins are cleaved to form their final active structures. Insulin, for example, is synthesized as preproinsulin, which yields proinsulin after the signal peptide has been cleaved. To form the mature insulin, the proinsulin is then cleaved at two positions to yield two polypeptide chains linked by 2 disulphide bonds. Proinsulin is necessary for the folding of the polypeptide chain, as the 2 polypeptide chains of insulin may not correctly assemble into the correct form, whereas its precursor proinsulin does.

Proteases in particular are synthesized in the inactive form so that they may be safely stored in cells, and ready for release in sufficient quantity when required. This is to ensure that the protease is activated only in the correct location or context, as inappropriate activation of these proteases can be very destructive for an organism. Proteolysis of the zymogen yields an active protein; for example, when trypsinogen is cleaved to form trypsin, a slight rearrangement of the protein structure that completes the active site of the protease occurs, thereby activating the protein.

Proteolysis can, therefore, be a method of regulating biological processes by turning inactive proteins into active ones. A good example is the blood clotting cascade whereby an initial event triggers a cascade of sequential proteolytic activation of many specific proteases, resulting in blood coagulation. The complement system of the immune response also involves a complex sequential proteolytic activation and interaction that result in an attack on invading pathogens.

Protein Degradation

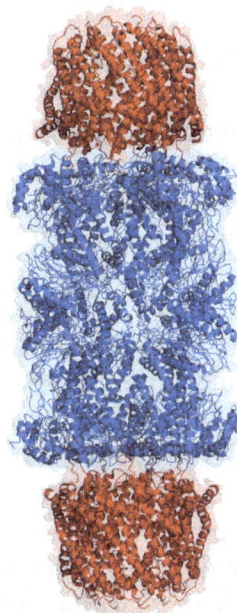

Structure of a proteasome. Its active sites are inside the tube (blue) where proteins are degraded.

Protein degradation may take place intracellularly or extracellularly. In digestion of food, digestive enzymes may be released into the environment for extracellular digestion whereby proteolytic cleavage breaks down proteins into smaller peptides and amino acids so that they may be absorbed and used by an organism. In animals the food may be processed extracellularly in specialized digestive organs or guts, but in many bacteria the food may be internalized into the cell via phagocytosis. Microbial degradation of protein in the environment can be regulated by nutrient availability. For example, limitation for major elements in proteins (carbon, nitrogen, and sulfur) has been shown to induce proteolytic activity in the fungus Neurospora crassa as well as in whole communities of soil organisms.

Proteins in cells are also constantly being broken down into amino acids. This intracellular degradation of protein serves a number of functions: It removes damaged and abnormal protein and prevent their accumulation, and it also serves to regulate cellular processes by removing enzymes and regulatory proteins that are no longer needed. The amino acids may then be reused for protein synthesis.

Lysosome and Proteasome

The intracellular degradation of protein may be achieved in two ways - proteolysis in lysosome, or a ubiquitin-dependent process that targets unwanted proteins to proteasome. The autophagy-lysosomal pathway is normally a non-selective process, but it may become selective upon starvation whereby proteins with peptide sequence KFERQ or similar are selectively broken down. The lysosome contains a large number of proteases such as cathepsins.

The ubiquitin-mediated process is selective. Proteins marked for degradation are covalently linked to ubiquitin. Many molecules of ubiquitin may be linked in tandem to a protein destined for degradation. The polyubiquinated protein is targeted to an ATP-dependent protease complex, the proteasome. The ubiquitin is released and reused, while the targeted protein is degraded.

Rate of Intracellular Protein Degradation

Different proteins are degraded at different rates. Abnormal proteins are quickly degraded, whereas the rate of degradation of normal proteins may vary widely depending on their functions. Enzymes at important metabolic control points may be degraded much faster than those enzymes whose activity is largely constant under all physiological conditions. One of the most rapidly degraded proteins is ornithine decarboxylase, which has a half-life of 11 minutes. In contrast, other proteins like actin and myosin have a half-life of a month or more, while, in essence, haemoglobin lasts for the entire life-time of an erythrocyte.

The N-end rule may partially determine the half-life of a protein, and proteins with segments rich in proline, glutamic acid, serine, and threonine (the so-called PEST proteins) have short half-life. Other factors suspected to affect degradation rate include the rate deamination of glutamine and asparagine and oxidation of cystein, histidine, and methionine, the absence of stabilizing ligands, the presence of attached carbohydrate or phosphate groups, the presence of free α-amino group, the negative charge of protein, and the flexibility and stability of the protein. Proteins with larger degrees of intrinsic disorder also tend to have short cellular half-life, with disordered segments having been proposed to facilitate efficient initiation of degradation by the proteasome.

The rate of proteolysis may also depend on the physiological state of the organism, such as its hormonal state as well as nutritional status. In time of starvation, the rate of protein degradation increases.

Digestion

In human digestion, proteins in food are broken down into smaller peptide chains by digestive enzymes such as pepsin, trypsin, chymotrypsin, and elastase, and into amino acids by various enzymes such as carboxypeptidase, aminopeptidase, and dipeptidase. It is necessary to break down proteins into small peptides (tripeptides and dipeptides) and amino acids so they can be absorbed by the intestines, and the absorbed tripeptides and dipeptides are also further broken into amino acids intracellularly before they enter the bloodstream. Different enzymes have different specificity for their substrate; trypsin, for example, cleaves the peptide bond after a positively charged residue (arginine and lysine); chymotrypsin cleaves the bond after an aromatic residue (phenylalanine, tyrosine, and tryptophan); elastase cleaves the bond after a small non-polar residue such as alanine or glycine.

In order to prevent inappropriate or premature activation of the digestive enzymes (they may, for example, trigger pancreatic self-digestion causing pancreatitis), these enzymes are secreted as inactive zymogen. The precursor of pepsin, pepsinogen, is secreted by the stomach, and is activated only in the acidic environment found in stomach. The pancreas secretes the precursors of a number of proteases such as trypsin and chymotrypsin. The zymogen of trypsin is trypsinogen, which is activated by a very specific protease, enterokinase, secreted by the mucosa of the duodenum. The trypsin, once activated, can also cleave other trypsinogens as well as the precursors of other proteases such as chymotrypsin and carboxypeptidase to activate them.

In bacteria, a similar strategy of employing an inactive zymogen or prezymogen is used. Subtilisin, which is produced by Bacillus subtilis, is produced as preprosubtilisin, and is released only if the signal peptide is cleaved and autocatalytic proteolytic activation has occurred.

Cellular Regulation

Proteolysis is also involved in the regulation of many cellular processes by activating or deactivating enzymes, transcription factors, and receptors, for example in the biosynthesis of cholesterol, or the mediation of thrombin signalling through protease-activated receptors.

Some enzymes at important metabolic control points such as ornithine decarboxylase is regulated entirely by its rate of synthesis and its rate of degradation. Other rapidly degraded proteins include the protein products of proto-oncogenes, which play central roles in the regulation of cell growth.

Cell cycle Regulation

Cyclins are a group of proteins that activate kinases involved in cell division. The degradation of cyclins is the key step that governs the exit from mitosis and progress into the next cell cycle. Cyclins accumulate in the course the cell cycle, then abruptly disappear just before the anaphase of mitosis. The cyclins are removed via a ubiquitin-mediated proteolytic pathway.

Apoptosis

Caspases are an important group of proteases involved in apoptosis. The precursors of caspase, procaspase, may be activated by proteolysis through its association with a protein complex that forms apoptosome, or by granzyme B, or via the death receptor pathways.

Proteolysis and Diseases

Abnormal proteolytic activity is associated with many diseases. In pancreatitis, leakage of proteases and their premature activation in the pancreas results in the self-digestion of the pancreas. People with diabetes mellitus may have increased lysosomal activity and the degradation of some proteins can increase significantly. Chronic inflammatory diseases such as rheumatoid arthritis may involve the release of lysosomal enzymes into extracellular space that break down surrounding tissues. Abnormal proteolysis and generation of peptides that aggregate in cells and their ineffective removal may result in many age-related neurological diseases such as Alzheimer's.

Proteases may be regulated by antiproteases or protease inhibitors, and imbalance between proteases and antiproteases can result in diseases, for example, in the destruction of lung tissues in emphysema brought on by smoking tobacco. Smoking is thought to increase the neutrophils and macrophages in the lung which release excessive amount of proteolytic enzymes such as elastase, such that they can no longer be inhibited by serpins such as α_1-antitrypsin, thereby resulting in the breaking down of connective tissues in the lung. Other proteases and their inhibitors may also be involved in this disease, for example matrix metalloproteinases (MMPs) and tissue inhibitors of metalloproteinases (TIMPs).

Other diseases linked to aberrant proteolysis include muscular dystrophy, degenerative skin disorders, respiratory and gastrointestinal diseases, and malignancy.

Non-Enzymatic Proteolysis

Protein backbones are very stable in water at neutral pH and room temperature, although the rate of hydrolysis of different peptide bonds can vary. The half life of a peptide bond under normal conditions can range from 7 years to 350 years, even higher for peptides protected by modified terminus or within the protein interior. The rate of proteolysis however can be significantly increased by extremes of pH and heat.

Strong mineral acids can readily hydrolyse the peptide bonds in a protein (acid hydrolysis). The standard way to hydrolyze a protein or peptide into its constituent amino acids for analysis is to heat it to 105 °C for around 24 hours in 6M hydrochloric acid. However, some proteins are resistant to acid hydrolysis. One well-known example is ribonuclease A, which can be purified by treating crude extracts with hot sulphuric acid so that other proteins become degraded while ribonuclease A is left intact.

Certain chemicals cause proteolysis only after specific residues, and these can be used to selectively break down a protein into smaller polypeptides for laboratory analysis. For example, cyanogen bromide cleaves the peptide bond after a methionine. Similar methods may be used to specifically cleave tryptophanyl, aspartyl, cysteinyl, and asparaginyl peptide bonds. Acids such as trifluoroacetic acid and formic acid may also be used.

Like other biomolecules, proteins can also be broken down by high heat alone. At 250 °C, the peptide bond may be easily hydrolyzed, with its half-life dropping to about a minute. Protein may also be broken down without hydrolysis through pyrolysis; small heterocyclic compounds may start to form upon degradation, above 500 °C, polycyclic aromatic hydrocarbon may also form, which is of interest in the study of generation of carcinogens in tobacco smoke and cooking at high heat.

Laboratory Applications

Proteolysis is also used in research and diagnostic applications:

- Cleavage of fusion protein so that the fusion partner and protein tag used in protein expression and purification may be removed. The proteases used have high degree of specificity, such as thrombin, enterokinase, and TEV protease, so that only the targeted sequence may be cleaved.

- Complete inactivation of undesirable enzymatic activity or removal of unwanted proteins. For example, proteinase K, a broad-spectrum proteinase stable in urea and SDS, is often used in the preparation of nucleic acids to remove unwanted nuclease contaminants that may otherwise degrade the DNA or RNA.

- Partial inactivation, or changing the functionality, of specific protein. For example, treatment of DNA polymerase I with subtilisin yields the Klenow fragment, which retains its polymerase function but lacks 5'-exonuclease activity.

- Digestion of proteins in solution for proteome analysis by liquid chromatography-mass spectrometry (LC-MS). This may also be done by in-gel digestion of proteins after separation by gel electrophoresis for the identification by mass spectrometry.

- Analysis of the stability of folded domain under a wide range of conditions.

- Increasing success rate of crystallisation projects

- Production of digested protein used in growth media to culture bacteria and other organisms, e.g. tryptone in Lysogeny Broth.

Protease Enzymes

Proteases may be classified according to the catalytic group involved in its active site.

- Cysteine protease

- Serine protease

- Threonine protease

- Aspartic protease

- Glutamic protease

- Metalloprotease

- Asparagine peptide lyase

Venoms

Certain types of venom, such as those produced by venomous snakes, can also cause proteolysis. These venoms are, in fact, complex digestive fluids that begin their work outside of the body. Proteolytic venoms cause a wide range of toxic effects, including effects that are:

- cytotoxic (cell-destroying)

- hemotoxic (blood-destroying)

- myotoxic (muscle-destroying)

- hemorrhagic (bleeding)

References

- Thomas E Creighton (1993). Proteins: Structures and Molecular Properties (2nd ed.). W H Freeman and Company. pp. 78–86. ISBN 0-7167-2317-4.

- Thomas E Creighton (1993). "Chapter 10 - Degradation". Proteins: Structures and Molecular Properties (2nd ed.). W H Freeman and Company. pp. 463–473. ISBN 0-7167-2317-4.

- Bernard Testa, Joachim M. Mayer (1 July 2003). Hydrolysis in Drug and Prodrug Metabolism. Wiley VCH. pp. 270–288. ISBN 978-3906390253.

- Thomas E Creighton (1993). Proteins: Structures and Molecular Properties (2nd ed.). W H Freeman and Company. p. 6. ISBN 0-7167-2317-4.

- Bryan John Smith (2002). "Chapter 71-75". In John M. Walker. The Protein Protocols Handbook (2 ed.). Humana Press. pp. 485–510. doi:10.1385/1592591698. ISBN 978-0-89603-940-7.

- Kohei Oda (2012). "New families of carboxyl peptidases: serine-carboxyl peptidases and glutamic peptidases". Journal of Biochemistry. 151 (1): 13–25. doi:10.1093/jb/mvr129. PMID 22016395.

- Inobe, Tomonao; Matouschek, Andreas (2014-02-01). "Paradigms of protein degradation by the proteasome". Current Opinion in Structural Biology. 24: 156–164. doi:10.1016/j.sbi.2014.02.002. ISSN 1879-033X. PMC 4010099. PMID 24632559.

- De Strooper B. (2010). "Proteases and proteolysis in Alzheimer disease: a multifactorial view on the disease process". Physiological Reviews. 90 (2): 465–94. doi:10.1152/physrev.00023.2009. PMID 20393191

- Fabbri D, Adamiano A, Torri C (2010). "GC-MS determination of polycyclic aromatic hydrocarbons evolved from pyrolysis of biomass". Anal Bioanal Chem. 397 (1): 309–17. doi:10.1007/s00216-010-3563-5. PMID 20213167.

Genetic Code and Metabolism: An Overview

Metabolism plays an important role in biochemistry. Many experiments and theories in biochemistry are conducted by studying metabolism. This chapter will help students understand the simple as well as complex concepts of metabolism in biochemistry. It will also provide information about the genetic code and what it is made of.

Genetic Code

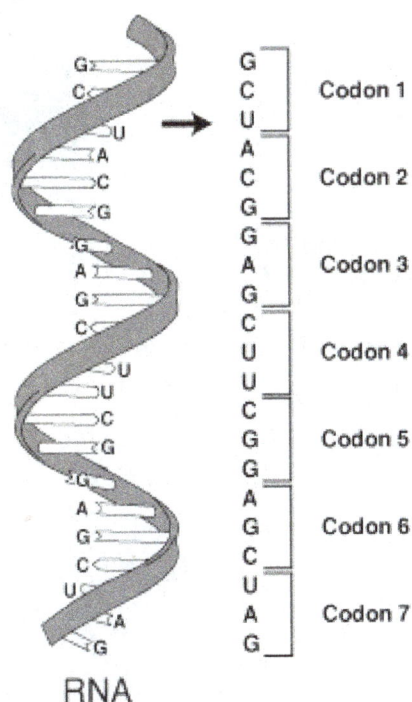

G
C   Codon 1
U

A
C   Codon 2
G

G
A   Codon 3
G

C
U   Codon 4
U

C
G   Codon 5
G

A
G   Codon 6
C

U
A   Codon 7
G

RNA

Ribonucleic acid

A series of codons in part of a messenger RNA (mRNA) molecule. Each codon consists of three nucleotides, usually corresponding to a single amino acid. The nucleotides are abbreviated with the letters A, U, G and C. This is mRNA, which uses U (uracil). DNA uses T (thymine) instead. This mRNA molecule will instruct a ribosome to synthesize a protein according to this code.

The genetic code is the set of rules by which information encoded within genetic material (DNA or mRNA sequences) is translated into proteins by living cells. Translation is accomplished by the ribosome, which links amino acids in an order specified by mRNA, using transfer RNA (tRNA) molecules to carry amino acids and to read the mRNA three nucleotides at a time. The genetic code is highly similar among all organisms and can be expressed in a simple table with 64 entries.

The code defines how sequences of nucleotide triplets, called codons, specify which amino acid will be added next during protein synthesis. With some exceptions, a three-nucleotide codon in a nucleic acid sequence specifies a single amino acid. Because the vast majority of genes are encoded with exactly the same code, this particular code is often referred to as the canonical or standard genetic code, or simply the genetic code, though in fact some variant codes have evolved. For example, protein synthesis in human mitochondria relies on a genetic code that differs from the standard genetic code.

While the "genetic code" determines a protein's amino acid sequence, other genomic regions determine when and where these proteins are produced according to a multitude of more complex "gene regulatory codes".

Discovery

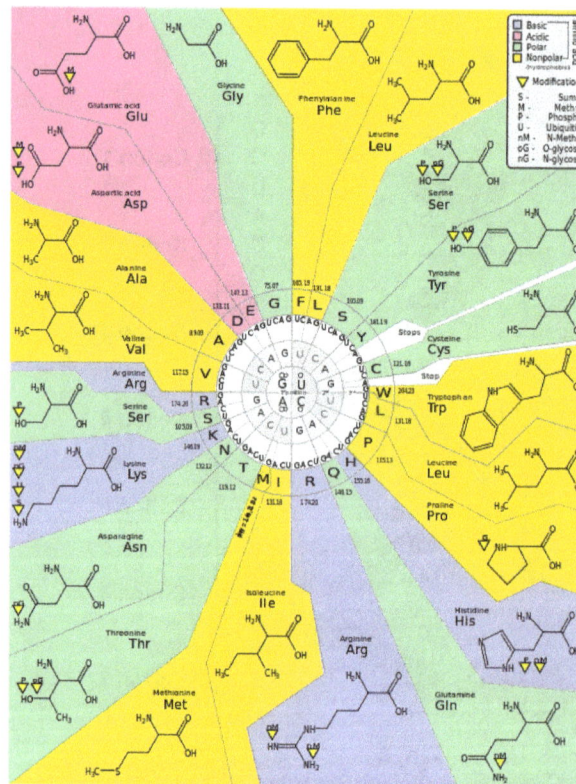

The genetic code

Serious efforts to understand how proteins are encoded began after the structure of DNA was discovered in 1953. George Gamow postulated that sets of three bases must be employed to encode the 20 standard amino acids used by living cells to build proteins. With four different nucleotides, a code of 2 nucleotides would allow for only a maximum of 4 = 16 amino acids. A code of 3 nucleotides could code for a maximum of 4 = 64 amino acids.

The Crick, Brenner et al. experiment first demonstrated that codons consist of three DNA bases; Marshall Nirenberg and Heinrich J. Matthaei were the first to elucidate the nature of a codon in 1961 at the National Institutes of Health. They used a cell-free system to translate a poly-uracil RNA sequence (i.e., UUUUU...) and discovered that the polypeptide that they had synthesized consisted of only the amino acid phenylalanine. They thereby deduced that the codon UUU specified the amino acid phenylalanine. This was followed by experiments in Severo Ochoa's laboratory that demonstrated that the poly-adenine RNA sequence (AAAAA...) coded for the polypeptide poly-lysine and that the poly-cytosine RNA sequence (CCCCC...) coded for the polypeptide poly-proline. Therefore, the codon AAA specified the amino acid lysine, and the codon CCC specified the amino acid proline. Using different copolymers most of the remaining codons were then determined. Subsequent work by Har Gobind Khorana identified the rest of the genetic code. Shortly thereafter, Robert W. Holley determined the structure of transfer RNA (tRNA), the adapter molecule that facilitates the process of translating RNA into protein. This work was based upon earlier studies by Severo Ochoa, who received the Nobel Prize in Physiology or Medicine in 1959 for his work on the enzymology of RNA synthesis.

Extending this work, Nirenberg and Philip Leder revealed the triplet nature of the genetic code and deciphered the codons of the standard genetic code. In these experiments, various combinations of mRNA were passed through a filter that contained ribosomes, the components of cells that translate RNA into protein. Unique triplets promoted the binding of specific tRNAs to the ribosome. Leder and Nirenberg were able to determine the sequences of 54 out of 64 codons in their experiments. In 1968, Khorana, Holley and Nirenberg received the Nobel Prize in Physiology or Medicine for their work.

Features

Reading Frame

A codon is defined by the initial nucleotide from which translation starts and sets the frame for a run of uninterrupted triplets, which is known as an "open reading frame" (ORF). For example, the string GGGAAACCC, if read from the first position, contains the codons GGG, AAA, and CCC; and, if read from the second position, it contains the codons GGA and AAC; if read starting from the third position, GAA and ACC. Every sequence can, thus, be read in its 5' → 3' direction in three reading frames, each of which will produce a different amino acid sequence (in the given example, Gly-Lys-Pro, Gly-Asn, or Glu-Thr, respectively). With double-stranded DNA, there are six possible reading frames, three in the forward orientation on one strand and three reverse on the opposite strand. The actual frame from which a protein sequence is translated is defined by a start codon, usually the first AUG codon in the mRNA sequence.

In eukaryotes, ORFs in exons are often interrupted by introns.

Start/stop Codons

Translation starts with a chain initiation codon or start codon. Unlike stop codons, the codon alone is not sufficient to begin the process. Nearby sequences such as the Shine-Dalgarno sequence in E. coli and initiation factors are also required to start translation. The most common start codon is AUG, which is read as methionine or, in bacteria, as formylmethionine. Alternative start codons depending on the organism include "GUG" or "UUG"; these codons normally represent valine and leucine, respectively, but as start codons they are translated as methionine or formylmethionine.

The three stop codons have been given names: UAG is amber, UGA is opal (sometimes also called umber), and UAA is ochre. "Amber" was named by discoverers Richard Epstein and Charles Steinberg after their friend Harris Bernstein, whose last name means "amber" in German. The other two stop codons were named "ochre" and "opal" in order to keep the "color names" theme. Stop codons are also called "termination" or "nonsense" codons. They signal release of the nascent polypeptide from the ribosome because there is no cognate tRNA that has anticodons complementary to these stop signals, and so a release factor binds to the ribosome instead.

Effect of Mutations

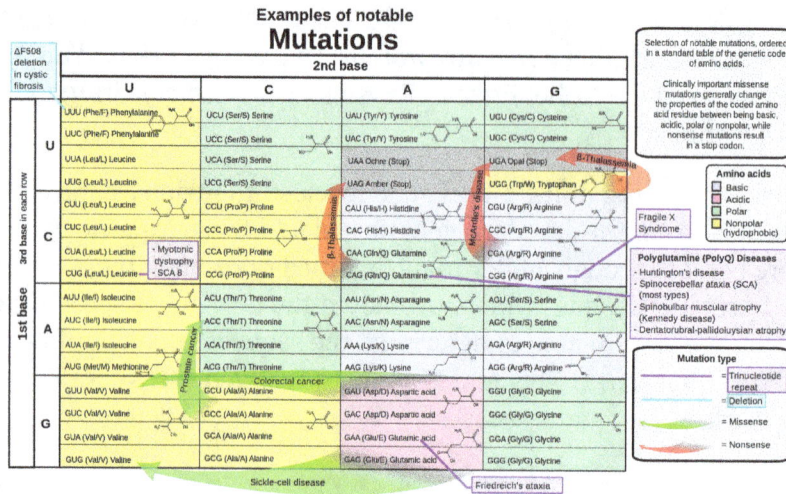

Examples of notable mutations that can occur in humans.

During the process of DNA replication, errors occasionally occur in the polymerization of the second strand. These errors, called mutations, can affect the phenotype of an organism, especially if they occur within the protein coding sequence of a gene. Error rates are usually very low—1 error in every 10–100 million bases—due to the "proofreading" ability of DNA polymerases.

Missense mutations and nonsense mutations are examples of point mutations, which can cause genetic diseases such as sickle-cell disease and thalassemia respectively. Clinically important missense mutations generally change the properties of the coded amino acid residue between being basic, acidic, polar or non-polar, whereas nonsense mutations result in a stop codon.

Mutations that disrupt the reading frame sequence by indels (insertions or deletions) of a non-multiple of 3 nucleotide bases are known as frameshift mutations. These mutations usually result in a completely different translation from the original, and are also very likely to cause a stop codon to be read, which truncates the creation of the protein. These mutations may impair the function of the resulting protein, and are thus rare in in vivo protein-coding sequences. One reason inheritance of frameshift mutations is rare is that, if the protein being translated is essential for growth under the selective pressures the organism faces, absence of a functional protein may cause death before the organism is viable. Frameshift mutations may result in severe genetic diseases such as Tay-Sachs disease.

Although most mutations that change protein sequences are harmful or neutral, some mutations have a beneficial effect on an organism. These mutations may enable the mutant organism to withstand particular environmental stresses better than wild type organisms, or reproduce more quickly. In these cases a mutation will tend to become more common in a population through natural selection. Viruses that use RNA as their genetic material have rapid mutation rates, which can be an advantage, since these viruses will evolve constantly and rapidly, and thus evade the defensive responses of e.g. the human immune system. In large populations of asexually reproducing organisms, for example, E. coli, multiple beneficial mutations may co-occur. This phenomenon is called clonal interference and causes competition among the mutations.

Degeneracy

Degeneracy is the redundancy of the genetic code. This term was given by Bernfield and Nirenberg. The genetic code has redundancy but no ambiguity. For example, although codons GAA and GAG both specify glutamic acid (redundancy), neither of them specifies any other amino acid (no ambiguity). The codons encoding one amino acid may differ in any of their three positions. For example, the amino acid leucine is specified by YUR or CUN (UUA, UUG, CUU, CUC, CUA, or CUG) codons (difference in the first or third position indicated using IUPAC notation), while the amino acid serine is specified by UCN or AGY (UCA, UCG, UCC, UCU, AGU, or AGC) codons (difference in the first, second, or third position). - - A practical consequence of redundancy is that errors in the third position of the triplet codon cause only a silent mutation or an error that would not affect the protein because the hydrophilicity or hydrophobicity is maintained by equivalent substitution of amino acids; for example, a codon of NUN (where N = any nucleotide) tends to code for hydrophobic amino acids. NCN yields amino acid residues that are small in size and moderate in hydropathy; NAN encodes average size hydrophilic residues. The genetic code is so well-structured for hydropathy that a mathematical analysis (Singular Value Decomposition) of 12 variables (4 nucleotides x 3 positions) yields a remarkable correlation (C = 0.95) for predicting the hydropathy of the encoded amino acid directly from the triplet nucleotide sequence, without translation. Note in the table, below, eight amino acids are not affected at all by mutations at the third position of the codon, whereas in the figure above, a mutation at the second position is likely to cause a radical change in the physicochemical properties of the encoded amino acid.

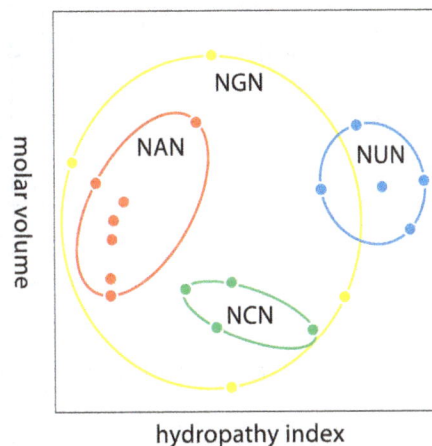

Grouping of codons by amino acid residue molar volume and hydropathy. A more detailed version is available.

Codon usage Bias

The frequency of codons, also known as "codon usage bias", can vary from species to species with functional implications for the control of translation. The following codon usage table is for the human genome.

Human genome codon frequency:

Codon	AA	Fraction	Freq ‰	Number	Codon	AA	Fraction	Freq ‰	Number	Codon	AA	Fraction	Freq ‰	Number	Codon	AA	Fraction	Freq ‰	Number
UUU	F	0.46	17.6	714298	UCU	S	0.19	15.2	618711	UAU	Y	0.44	12.2	495699	UGU	C	0.46	10.6	430311
UUC	F	0.54	20.3	824692	UCC	S	0.22	17.7	718892	UAC	Y	0.56	15.3	622407	UGC	C	0.54	12.6	513028
UUA	L	0.08	7.7	311881	UCA	S	0.15	12.2	496448	UAA	*	0.30	1.0	40285	UGA	*	0.47	1.6	63237
UUG	L	0.13	12.9	525688	UCG	S	0.05	4.4	179419	UAG	*	0.24	0.8	32109	UGG	W	1.00	13.2	535595
CUU	L	0.13	13.2	536515	CCU	P	0.29	17.5	713233	CAU	H	0.42	10.9	441711	CGU	R	0.08	4.5	184609
CUC	L	0.20	19.6	796638	CCC	P	0.32	19.8	804620	CAC	H	0.58	15.1	613713	CGC	R	0.18	10.4	423516
CUA	L	0.07	7.2	290751	CCA	P	0.28	16.9	688038	CAA	Q	0.27	12.3	501911	CGA	R	0.11	6.2	250760
CUG	L	0.40	39.6	1611801	CCG	P	0.11	6.9	281570	CAG	Q	0.73	34.2	1391973	CGG	R	0.20	11.4	464485
AUU	I	0.36	16.0	650473	ACU	T	0.25	13.1	533609	AAU	N	0.47	17.0	689701	AGU	S	0.15	12.1	493429
AUC	I	0.47	20.8	846466	ACC	T	0.36	18.9	768147	AAC	N	0.53	19.1	776603	AGC	S	0.24	19.5	791383
AUA	I	0.17	7.5	304565	ACA	T	0.28	15.1	614523	AAA	K	0.43	24.4	993621	AGA	R	0.21	12.2	494682
AUG	M	1.00	22.0	896005	ACG	T	0.11	6.1	246105	AAG	K	0.57	31.9	1295568	AGG	R	0.21	12.0	486463
GUU	V	0.18	11.0	448607	GCU	A	0.27	18.4	750096	GAU	D	0.46	21.8	885429	GGU	G	0.16	10.8	437126
GUC	V	0.24	14.5	588138	GCC	A	0.40	27.7	1127679	GAC	D	0.54	25.1	1020595	GGC	G	0.34	22.2	903565
GUA	V	0.12	7.1	287712	GCA	A	0.23	15.8	643471	GAA	E	0.42	29.0	1177632	GGA	G	0.25	16.5	669873
GUG	V	0.46	28.1	1143534	GCG	A	0.11	7.4	299495	GAG	E	0.58	39.6	1609975	GGG	G	0.25	16.5	669768

Variation

While slight variations on the standard code had been predicted earlier, none were discovered until 1979, when researchers studying human mitochondrial genes discovered they used an alternative code. Many slight variants have been discovered since then, including various alternative mitochondrial codes, and small variants such as translation of the codon UGA as tryptophan in Mycoplasma species, and translation of CUG as a serine rather than a leucine in yeasts of the "CTG clade" (Candida albicans is member of this group). Because viruses must use the same genetic code as their hosts, modifications to the standard genetic code could interfere with the synthesis or functioning of viral proteins. However, some viruses (such as totiviruses) have adapted to the genetic code modification of the host. In bacteria and archaea, GUG and UUG are common start codons, but in rare cases, certain proteins may use alternative start codons not normally used by that species.

In certain proteins, non-standard amino acids are substituted for standard stop codons, depending on associated signal sequences in the messenger RNA. For example, UGA can code for selenocysteine and UAG can code for pyrrolysine. Selenocysteine is now viewed as the 21st amino acid, and pyrrolysine is viewed as the 22nd. Unlike selenocysteine, pyrrolysine encoded UAG is translated with the participation of a dedicated aminoacyl-tRNA synthetase. Both selenocysteine and pyrrolysine may be present in the same organism. Although the genetic code is normally fixed in an organism, the achaeal prokaryote Acetohalobium arabaticum can expand its genetic code from 20 to 21 amino acids (by including pyrrolysine) under different conditions of growth.

Despite these differences, all known naturally-occurring codes are very similar to each other, and the coding mechanism is the same for all organisms: three-base codons, tRNA, ribosomes, reading the code in the same direction and translating the code three letters at a time into sequences of amino acids.

Genetic code logo of the Globobulimina pseudospinescens mitochondrial genome. The logo shows the 64 codons from left to right, predicted alternatives in red (relative to the standard genetic code). Red line: stop codons. The height of each amino acid in the stack shows how often it is aligned to the codon in homologous protein domains. The stack height indicates the support for the prediction.

Variant genetic codes used by an organism can be inferred by identifying highly conserved genes encoded in that genome, and comparing its codon usage to the amino acids in homologous proteins of other organisms. For example, the program FACIL infers a genetic code by searching which amino acids in homologous protein domains are most often aligned to every codon. The resulting amino acid probabilities for each codon are displayed in a genetic code logo, that also shows the support for a stop codon.

RNA Codon Table

nonpolar	polar	basic	acidic	(stop codon)

Standard genetic code									
1st base	2nd base								3rd base
	U		C		A		G		
U	UUU	(Phe/F) Phenylala-nine	UCU	(Ser/S) Serine	UAU	(Tyr/Y) Tyrosine	UGU	(Cys/C) Cysteine	U
	UUC		UCC		UAC		UGC		C
	UUA	(Leu/L) Leucine	UCA		UAA	Stop (Ochre)	UGA	Stop (Opal)	A
	UUG		UCG		UAG	Stop (Amber)	UGG	(Trp/W) Tryptophan	G
C	CUU		CCU	(Pro/P) Proline	CAU	(His/H) Histidine	CGU	(Arg/R) Arginine	U
	CUC		CCC		CAC		CGC		C
	CUA		CCA		CAA	(Gln/Q) Glutamine	CGA		A
	CUG		CCG		CAG		CGG		G
A	AUU	(Ile/I) Isoleucine	ACU	(Thr/T) Threonine	AAU	(Asn/N) Asparagine	AGU	(Ser/S) Serine	U
	AUC		ACC		AAC		AGC		C
	AUA		ACA		AAA	(Lys/K) Lysine	AGA	(Arg/R) Arginine	A
	AUG[A]	(Met/M) Methionine	ACG		AAG		AGG		G
G	GUU	(Val/V) Valine	GCU	(Ala/A) Alanine	GAU	(Asp/D) Aspartic acid	GGU	(Gly/G) Glycine	U
	GUC		GCC		GAC		GGC		C
	GUA		GCA		GAA	(Glu/E) Glutamic acid	GGA		A
	GUG		GCG		GAG		GGG		G

The codon AUG both codes for methionine and serves as an initiation site: the first AUG in an mRNA's coding region is where translation into protein begins.

Inverse table (compressed using IUPAC notation)					
Amino acid	Codons	Compressed	Amino acid	Codons	Compressed
Ala/A	GCU, GCC, GCA, GCG	GCN	Leu/L	UUA, UUG, CUU, CUC, CUA, CUG	YUR, CUN
Arg/R	CGU, CGC, CGA, CGG, AGA, AGG	CGN, MGR	Lys/K	AAA, AAG	AAR
Asn/N	AAU, AAC	AAY	Met/M	AUG	
Asp/D	GAU, GAC	GAY	Phe/F	UUU, UUC	UUY
Cys/C	UGU, UGC	UGY	Pro/P	CCU, CCC, CCA, CCG	CCN
Gln/Q	CAA, CAG	CAR	Ser/S	UCU, UCC, UCA, UCG, AGU, AGC	UCN, AGY
Glu/E	GAA, GAG	GAR	Thr/T	ACU, ACC, ACA, ACG	ACN
Gly/G	GGU, GGC, GGA, GGG	GGN	Trp/W	UGG	
His/H	CAU, CAC	CAY	Tyr/Y	UAU, UAC	UAY
Ile/I	AUU, AUC, AUA	AUH	Val/V	GUU, GUC, GUA, GUG	GUN
START	AUG		STOP	UAA, UGA, UAG	UAR, URA

DNA Codon Table

The DNA codon table is essentially identical to that for RNA, but with U replaced by T.

Origin

The origin of the genetic code is a part of the question of the origin of life. Under the main hypothesis for the origin of life, the RNA world hypothesis, any model for the emergence of genetic code is intimately related to a model of the transfer from ribozymes (RNA enzymes) to proteins as the principal enzymes in cells. In line with the RNA world hypothesis, transfer RNA molecules appear to have evolved before modern aminoacyl-tRNA synthetases, so the latter cannot be part of the explanation of its patterns.

A consideration of a hypothetical random genetic code further motivates a biochemical or evolutionary model for the origin of the genetic code. If amino acids were randomly assigned to triplet codons, there would be 1.5×10 possible genetic codes to choose from. This number is found by calculating how many ways there are to place 21 items (20 amino acids plus one stop) in 64 bins, wherein each item is used at least once. In fact, the distribution of codon assignments in the genetic code is nonrandom. In particular, the genetic code clusters certain amino acid assignments. For example, amino acids that share the same biosynthetic pathway tend to have the same first base in their codons. This could be an evolutionary relic of early simpler genetic code with fewer amino acids, that later diverged to code for a larger set of amino acids. It could also reflect steric and chemical properties that had and other effect on the codon during its evolution. Amino acids with similar physical properties also tend to have similar codons, reducing the problems caused by point mutations and mistranslations.

Given the non-random genetic triplet coding scheme, it has been suggested that a tenable hypothesis for the origin of genetic code should address multiple aspects of the codon table such as absence of codons for D-amino acids, secondary codon patterns for some amino acids, secondary codon patterns for some amino acids, confinement of synonymous positions to third position, a limited set of only 20 amino acids instead of a number closer to 64, and the relation of stop codon patterns to amino acid coding patterns.

There are three main ideas for the origin of the genetic code, and many models belong to either one of them or to a combination thereof:

1. "Frozen accident": the genetic code has been randomly created. For example, early tRNA-like ribozymes may have had different affinities for amino acids, with codons emerging from another part of the ribozyme which exhibited random variability. Once enough peptides were coded for, any major random change in the genetic code would have been lethal, hence it is "frozen".

2. Stereochemical affinity: the genetic code is a result of a high affinity between each amino acid and its codon or anti-codon; the latter option implies that pre-tRNA molecules matched their corresponding amino acids by this affinity. Later during evolution, this matching has been gradually replaced with the one performed today by aminoacyl-tRNA synthetases.

3. Optimality: the genetic code continued to evolve after its initial creation, so that the current

code reflects adaptation that maximizes some fitness function, usually some kind of error minimization.

Hypotheses for the origin of the genetic code have addressed a variety of scenarios:

- Chemical principles govern specific RNA interaction with amino acids. Experiments with aptamers showed that some amino acids have a selective chemical affinity for the base triplets that code for them. Recent experiments show that of the 8 amino acids tested, 6 show some RNA triplet-amino acid association.

- Biosynthetic expansion. The standard modern genetic code grew from a simpler earlier code through a process of "biosynthetic expansion". Here the idea is that primordial life "discovered" new amino acids (for example, as by-products of metabolism) and later incorporated some of these into the machinery of genetic coding. Although much circumstantial evidence has been found to suggest that fewer different amino acids were used in the past than today, precise and detailed hypotheses about which amino acids entered the code in what order have proved far more controversial.

- Natural selection has led to codon assignments of the genetic code that minimize the effects of mutations. A recent hypothesis suggests that the triplet code was derived from codes that used longer than triplet codons (such as quadruplet codons). Longer than triplet decoding would have higher degree of codon redundancy and would be more error resistant than the triplet decoding. This feature could allow accurate decoding in the absence of highly complex translational machinery such as the ribosome and before cells began making ribosomes.

- Information channels: Information-theoretic approaches model the process of translating the genetic code into corresponding amino acids as an error-prone information channel. The inherent noise (that is, the error) in the channel poses the organism with a fundamental question: how can a genetic code be constructed to withstand the effect of noise while accurately and efficiently translating information? These "rate-distortion" models suggest that the genetic code originated as a result of the interplay of the three conflicting evolutionary forces: the needs for diverse amino-acids, for error-tolerance and for minimal cost of resources. The code emerges at a coding transition when the mapping of codons to amino-acids becomes nonrandom. The emergence of the code is governed by the topology defined by the probable errors and is related to the map coloring problem.

- Game theory: Models based on signaling games combine elements of game theory, natural selection and information channels. Such models have been used to suggest that the first polypeptides were likely short and had some use other than enzymatic function. Game theoretic models have also suggested that the organization of RNA strings into cells may have been necessary to prevent "deceptive" use of the genetic code, i.e. preventing the ancient equivalent of viruses from overwhelming the RNA world.

- Stop codons: Codons for translational stops are also an interesting aspect to the problem of the origin of the genetic code. As an example for addressing stop codon evolution, it has been suggested that the stop codons are such that they are most likely to terminate translation early in the case of a frame shift error. In contrast, some stereochemical molecular models explain the origin of stop codons as "unassignable".

Expanded Genetic Codes (Synthetic Biology)

Since 2001, 40 non-natural amino acids have been added into protein by creating a unique codon (recoding) and a corresponding transfer-RNA:aminoacyl – tRNA-synthetase pair to encode it with diverse physicochemical and biological properties in order to be used as a tool to exploring protein structure and function or to create novel or enhanced proteins.

H. Murakami and M. Sisido have extended some codons to have four and five bases. Steven A. Benner constructed a functional 65th (in vivo) codon.

Metabolism

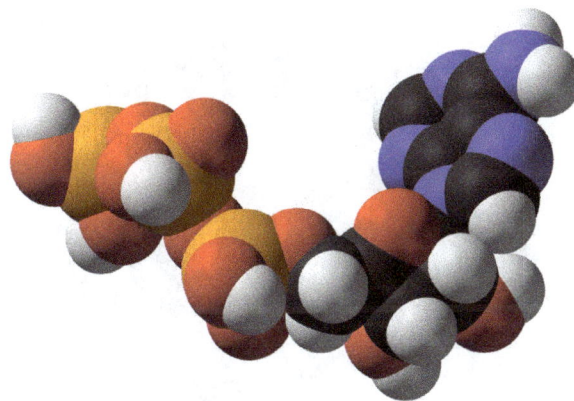

Structure of adenosine triphosphate (ATP), a central intermediate in energy metabolism

Metabolism is the set of life-sustaining chemical transformations within the cells of living organisms. The three main purposes of metabolism are the conversion of food/fuel to energy to run cellular processes, the conversion of food/fuel to building blocks for proteins, lipids, nucleic acids, and some carbohydrates, and the elimination of nitrogenous wastes. These enzyme-catalyzed reactions allow organisms to grow and reproduce, maintain their structures, and respond to their environments. The word metabolism can also refer to the sum of all chemical reactions that occur in living organisms, including digestion and the transport of substances into and between different cells, in which case the set of reactions within the cells is called intermediary metabolism or intermediate metabolism.

Metabolism is usually divided into two categories: catabolism, the breaking down of organic matter, for example, by cellular respiration, and anabolism, the building up of components of cells such as proteins and nucleic acids. Usually, breaking down releases energy and building up consumes energy.

The chemical reactions of metabolism are organized into metabolic pathways, in which one chemical is transformed through a series of steps into another chemical, by a sequence of enzymes. Enzymes are crucial to metabolism because they allow organisms to drive desirable reactions that require energy that will not occur by themselves, by coupling them to spontaneous reactions that release energy. Enzymes act as catalysts that allow the reactions to proceed more rapidly. Enzymes also allow the regulation of metabolic pathways in response to changes in the cell's environment or to signals from other cells.

The metabolic system of a particular organism determines which substances it will find nutritious and which poisonous. For example, some prokaryotes use hydrogen sulfide as a nutrient, yet this gas is poisonous to animals. The speed of metabolism, the metabolic rate, influences how much food an organism will require, and also affects how it is able to obtain that food.

A striking feature of metabolism is the similarity of the basic metabolic pathways and components between even vastly different species. For example, the set of carboxylic acids that are best known as the intermediates in the citric acid cycle are present in all known organisms, being found in species as diverse as the unicellular bacterium Escherichia coli and huge multicellular organisms like elephants. These striking similarities in metabolic pathways are likely due to their early appearance in evolutionary history, and their retention because of their efficacy.

Key Biochemicals

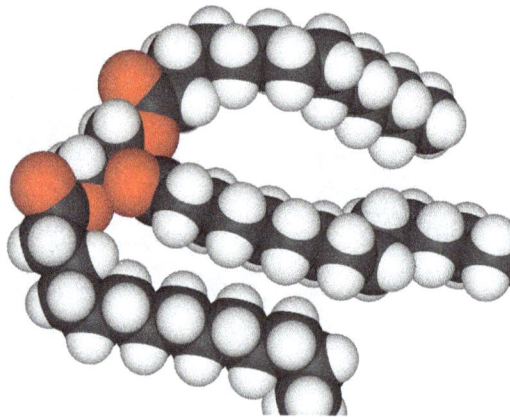

Structure of a triacylglycerol lipid

Most of the structures that make up animals, plants and microbes are made from three basic classes of molecule: amino acids, carbohydrates and lipids (often called fats). As these molecules are vital for life, metabolic reactions either focus on making these molecules during the construction of cells and tissues, or by breaking them down and using them as a source of energy, by their digestion. These biochemicals can be joined together to make polymers such as DNA and proteins, essential macromolecules of life.

Type of molecule	Name of monomer forms	Name of polymer forms	Examples of polymer forms
Amino acids	Amino acids	Proteins (made of polypeptides)	Fibrous proteins and globular proteins
Carbohydrates	Monosaccharides	Polysaccharides	Starch, glycogen and cellulose
Nucleic acids	Nucleotides	Polynucleotides	DNA and RNA

Amino Acids and Proteins

Proteins are made of amino acids arranged in a linear chain joined together by peptide bonds. Many proteins are enzymes that catalyze the chemical reactions in metabolism. Other proteins have structural or mechanical functions, such as those that form the cytoskeleton, a system of scaffolding that maintains the cell shape. Proteins are also important in cell signaling, immune responses, cell adhesion, active transport across membranes, and the cell cycle. Amino acids also contribute to cellular energy metabolism by providing a carbon source for entry into the citric acid cycle (tricarboxylic acid cycle), especially when a primary source of energy, such as glucose, is scarce, or when cells undergo metabolic stress.

Lipids

Lipids are the most diverse group of biochemicals. Their main structural uses are as part of biological membranes both internal and external, such as the cell membrane, or as a source of energy. Lipids are usually defined as hydrophobic or amphipathic biological molecules but will dissolve in organic solvents such as benzene or chloroform. The fats are a large group of compounds that contain fatty acids and glycerol; a glycerol molecule attached to three fatty acid esters is called a triacylglyceride. Several variations on this basic structure exist, including alternate backbones such as sphingosine in the sphingolipids, and hydrophilic groups such as phosphate as in phospholipids. Steroids such as cholesterol are another major class of lipids.

Carbohydrates

Carbohydrates are aldehydes or ketones, with many hydroxyl groups attached, that can exist as straight chains or rings. Carbohydrates are the most abundant biological molecules, and fill numerous roles, such as the storage and transport of energy (starch, glycogen) and structural components (cellulose in plants, chitin in animals). The basic carbohydrate units are called monosaccharides and include galactose, fructose, and most importantly glucose. Monosaccharides can be linked together to form polysaccharides in almost limitless ways.

Nucleotides

The two nucleic acids, DNA and RNA, are polymers of nucleotides. Each nucleotide is composed of a phosphate attached to a ribose or deoxyribose sugar group which is attached to a nitrogenous base. Nucleic acids are critical for the storage and use of genetic information, and its interpretation through the processes of transcription and protein biosynthesis. This information is protected by DNA repair mechanisms and propagated through DNA replication. Many viruses have an RNA genome, such as HIV, which uses reverse transcription to create a DNA template from its viral RNA genome. RNA in ribozymes such as spliceosomes and ribosomes is similar to enzymes as it can catalyze chemical reactions. Individual nucleosides are made by attaching a nucleobase to a ribose sugar. These bases are heterocyclic rings containing nitrogen, classified as purines or pyrimidines. Nucleotides also act as coenzymes in metabolic-group-transfer reactions.

Coenzymes

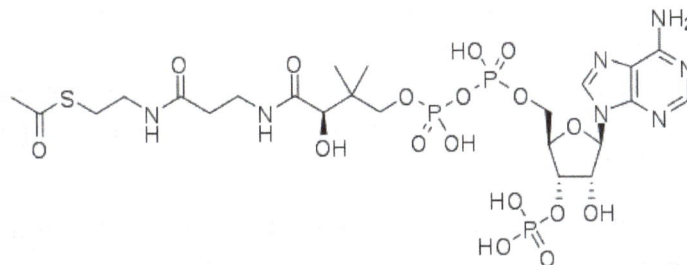

Structure of the coenzyme acetyl-CoA. The transferable acetyl group is bonded to the sulfur atom at the extreme left.

Metabolism involves a vast array of chemical reactions, but most fall under a few basic types of reactions that involve the transfer of functional groups of atoms and their bonds within molecules. This common chemistry allows cells to use a small set of metabolic intermediates to carry chemical groups between different reactions. These group-transfer intermediates are called coenzymes. Each class of group-transfer reactions is carried out by a particular coenzyme, which is the substrate for a set of enzymes that produce it, and a set of enzymes that consume it. These coenzymes are therefore continuously made, consumed and then recycled.

One central coenzyme is adenosine triphosphate (ATP), the universal energy currency of cells. This nucleotide is used to transfer chemical energy between different chemical reactions. There is only a small amount of ATP in cells, but as it is continuously regenerated, the human body can use about its own weight in ATP per day. ATP acts as a bridge between catabolism and anabolism. Catabolism breaks down molecules and anabolism puts them together. Catabolic reactions generate ATP and anabolic reactions consume it. It also serves as a carrier of phosphate groups in phosphorylation reactions.

A vitamin is an organic compound needed in small quantities that cannot be made in cells. In human nutrition, most vitamins function as coenzymes after modification; for example, all water-soluble vitamins are phosphorylated or are coupled to nucleotides when they are used in cells. Nicotinamide adenine dinucleotide (NAD^+), a derivative of vitamin B_3 (niacin), is an important coenzyme that acts as a hydrogen acceptor. Hundreds of separate types of dehydrogenases remove electrons from their substrates and reduce NAD^+ into NADH. This reduced form of the coenzyme is then a substrate for any of the reductases in the cell that need to reduce their substrates. Nicotinamide adenine dinucleotide exists in two related forms in the cell, NADH and NADPH. The NAD^+/NADH form is more important in catabolic reactions, while $NADP^+$/NADPH is used in anabolic reactions.

Minerals and Cofactors

Inorganic elements play critical roles in metabolism; some are abundant (e.g. sodium and potassium) while others function at minute concentrations. About 99% of a mammal's mass is made up of the elements carbon, nitrogen, calcium, sodium, chlorine, potassium, hydrogen, phosphorus, oxygen and sulfur. Organic compounds (proteins, lipids and carbohydrates) contain the majority of the carbon and nitrogen; most of the oxygen and hydrogen is present as water.

The abundant inorganic elements act as ionic electrolytes. The most important ions are sodium, potassium, calcium, magnesium, chloride, phosphate and the organic ion bicarbonate. The maintenance of precise ion gradients across cell membranes maintains osmotic pressure and pH. Ions are also critical for nerve and muscle function, as action potentials in these tissues are produced by the exchange of electrolytes between the extracellular fluid and the cell's fluid, the cytosol. Electrolytes enter and leave cells through proteins in the cell membrane called ion channels. For example, muscle contraction depends upon the movement of calcium, sodium and potassium through ion channels in the cell membrane and T-tubules.

Transition metals are usually present as trace elements in organisms, with zinc and iron being most abundant of those. These metals are used in some proteins as cofactors and are essential for the activity of enzymes such as catalase and oxygen-carrier proteins such as hemoglobin. Metal cofactors are bound tightly to specific sites in proteins; although enzyme cofactors can be modified during catalysis, they always return to their original state by the end of the reaction catalyzed. Metal micronutrients are taken up into organisms by specific transporters and bind to storage proteins such as ferritin or metallothionein when not in use.

Catabolism

Catabolism is the set of metabolic processes that break down large molecules. These include breaking down and oxidizing food molecules. The purpose of the catabolic reactions is to provide the energy and components needed by anabolic reactions. The exact nature of these catabolic reactions differ from organism to organism and organisms can be classified based on their sources of energy and carbon (their primary nutritional groups), as shown in the table below. Organic molecules are used as a source of energy by organotrophs, while lithotrophs use inorganic substrates and phototrophs capture sunlight as chemical energy. However, all these different forms of metabolism depend on redox reactions that involve the transfer of electrons from reduced donor molecules such as organic molecules, water, ammonia, hydrogen sulfide or ferrous ions to acceptor molecules such as oxygen, nitrate or sulfate. In animals these reactions involve complex organic molecules that are broken down to simpler molecules, such as carbon dioxide and water. In photosynthetic organisms such as plants and cyanobacteria, these electron-transfer reactions do not release energy, but are used as a way of storing energy absorbed from sunlight.

Classification of organisms based on their metabolism						
Energy source	sunlight	photo-				-troph
	Preformed molecules	chemo-				
Electron donor	organic compound			organo-		
	inorganic compound			litho-		
Carbon source	organic compound				hetero-	
	inorganic compound	auto-				

The most common set of catabolic reactions in animals can be separated into three main stages. In the first, large organic molecules such as proteins, polysaccharides or lipids are digested into their smaller components outside cells. Next, these smaller molecules are taken up by cells and converted to yet smaller molecules, usually acetyl coenzyme A (acetyl-CoA), which releases some energy. Finally, the acetyl group on the CoA is oxidised to water and carbon dioxide in the citric acid cycle and electron transport chain, releasing the energy that is stored by reducing the coenzyme nicotinamide adenine dinucleotide (NAD^+) into NADH.

Digestion

Macromolecules such as starch, cellulose or proteins cannot be rapidly taken up by cells and must be broken into their smaller units before they can be used in cell metabolism. Several common classes of enzymes digest these polymers. These digestive enzymes include proteases that digest proteins into amino acids, as well as glycoside hydrolases that digest polysaccharides into simple sugars known as monosaccharides.

Microbes simply secrete digestive enzymes into their surroundings, while animals only secrete these enzymes from specialized cells in their guts. The amino acids or sugars released by these extracellular enzymes are then pumped into cells by active transport proteins.

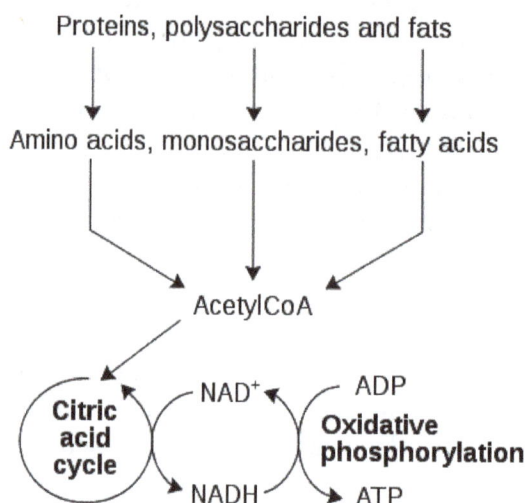

A simplified outline of the catabolism of proteins, carbohydrates and fats

Energy from Organic Compounds

Carbohydrate catabolism is the breakdown of carbohydrates into smaller units. Carbohydrates are usually taken into cells once they have been digested into monosaccharides. Once inside, the major route of breakdown is glycolysis, where sugars such as glucose and fructose are converted into pyruvate and some ATP is generated. Pyruvate is an intermediate in several metabolic pathways, but the majority is converted to acetyl-CoA and fed into the citric acid cycle. Although some more ATP is generated in the citric acid cycle, the most important product is NADH, which is made from NAD^+ as the acetyl-CoA is oxidized. This oxidation releases carbon dioxide as a waste product. In anaerobic conditions, glycolysis produces lactate, through the enzyme lactate dehydrogenase re-oxidizing NADH to NAD+ for re-use in glycolysis. An alternative route for glucose breakdown is the pentose phosphate pathway, which reduces the coenzyme NADPH and produces pentose sugars such as ribose, the sugar component of nucleic acids.

Fats are catabolised by hydrolysis to free fatty acids and glycerol. The glycerol enters glycolysis and the fatty acids are broken down by beta oxidation to release acetyl-CoA, which then is fed into the citric acid cycle. Fatty acids release more energy upon oxidation than carbohydrates because carbohydrates contain more oxygen in their structures. Steroids are also broken down by some bacteria in a process similar to beta oxidation, and this breakdown process involves the release of significant amounts of acetyl-CoA, propionyl-CoA, and pyruvate, which can all be used by the cell for energy. M. tuberculosis can also grow on the lipid cholesterol as a sole source of carbon, and genes involved in the cholesterol use pathway(s) have been validated as important during various stages of the infection lifecycle of M. tuberculosis.

Amino acids are either used to synthesize proteins and other biomolecules, or oxidized to urea and carbon dioxide as a source of energy. The oxidation pathway starts with the removal of the amino group by a transaminase. The amino group is fed into the urea cycle, leaving a deaminated carbon skeleton in the form of a keto acid. Several of these keto acids are intermediates in the citric acid cycle, for example the deamination of glutamate forms α-ketoglutarate. The glucogenic amino acids can also be converted into glucose, through gluconeogenesis (discussed below).

Energy Transformations

Oxidative Phosphorylation

In oxidative phosphorylation, the electrons removed from organic molecules in areas such as the protagon acid cycle are transferred to oxygen and the energy released is used to make ATP. This is done in eukaryotes by a series of proteins in the membranes of mitochondria called the electron transport chain. In prokaryotes, these proteins are found in the cell's inner membrane. These proteins use the energy released from passing electrons from reduced molecules like NADH onto oxygen to pump protons across a membrane.

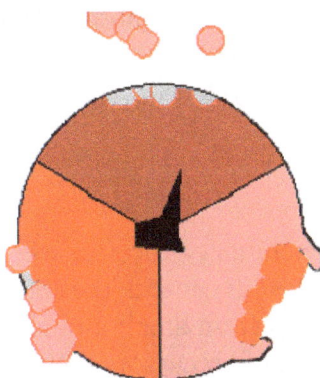

Mechanism of ATP synthase. ATP is shown in red, ADP and phosphate in pink and the rotating stalk subunit in black.

Pumping protons out of the mitochondria creates a proton concentration difference across the membrane and generates an electrochemical gradient. This force drives protons back into the mitochondrion through the base of an enzyme called ATP synthase. The flow of protons makes the stalk subunit rotate, causing the active site of the synthase domain to change shape and phosphorylate adenosine diphosphate – turning it into ATP.

Energy from Inorganic Compounds

Chemolithotrophy is a type of metabolism found in prokaryotes where energy is obtained from the oxidation of inorganic compounds. These organisms can use hydrogen, reduced sulfur compounds (such as sulfide, hydrogen sulfide and thiosulfate), ferrous iron (FeII) or ammonia as sources of reducing power and they gain energy from the oxidation of these compounds with electron acceptors such as oxygen or nitrite. These microbial processes are important in global biogeochemical cycles such as acetogenesis, nitrification and denitrification and are critical for soil fertility.

Energy from Light

The energy in sunlight is captured by plants, cyanobacteria, purple bacteria, green sulfur bacteria and some protists. This process is often coupled to the conversion of carbon dioxide into organic compounds, as part of photosynthesis, which is discussed below. The energy capture and carbon fixation systems can however operate separately in prokaryotes, as purple bacteria and green sulfur bacteria can use sunlight as a source of energy, while switching between carbon fixation and the fermentation of organic compounds.

In many organisms the capture of solar energy is similar in principle to oxidative phosphorylation, as it involves the storage of energy as a proton concentration gradient. This proton motive force then drives ATP synthesis. The electrons needed to drive this electron transport chain come from light-gathering proteins called photosynthetic reaction centres or rhodopsins. Reaction centers are classed into two types depending on the type of photosynthetic pigment present, with most photosynthetic bacteria only having one type, while plants and cyanobacteria have two.

In plants, algae, and cyanobacteria, photosystem II uses light energy to remove electrons from water, releasing oxygen as a waste product. The electrons then flow to the cytochrome b6f complex, which uses their energy to pump protons across the thylakoid membrane in the chloroplast. These protons move back through the membrane as they drive the ATP synthase, as before. The electrons then flow through photosystem I and can then either be used to reduce the coenzyme NADP$^+$, for use in the Calvin cycle, which is discussed below, or recycled for further ATP generation.

Anabolism

Anabolism is the set of constructive metabolic processes where the energy released by catabolism is used to synthesize complex molecules. In general, the complex molecules that make up cellular structures are constructed step-by-step from small and simple precursors. Anabolism involves three basic stages. First, the production of precursors such as amino acids, monosaccharides, isoprenoids and nucleotides, secondly, their activation into reactive forms using energy from ATP, and thirdly, the assembly of these precursors into complex molecules such as proteins, polysaccharides, lipids and nucleic acids.

Organisms differ in how many of the molecules in their cells they can construct for themselves. Autotrophs such as plants can construct the complex organic molecules in cells such as polysaccharides and proteins from simple molecules like carbon dioxide and water. Heterotrophs, on the other hand, require a source of more complex substances, such as monosaccharides and amino acids, to produce these complex molecules. Organisms can be further classified by ultimate source of their energy: photoautotrophs and photoheterotrophs obtain energy from light, whereas chemoautotrophs and chemoheterotrophs obtain energy from inorganic oxidation reactions.

Carbon Fixation

Photosynthesis is the synthesis of carbohydrates from sunlight and carbon dioxide (CO_2). In plants, cyanobacteria and algae, oxygenic photosynthesis splits water, with oxygen produced as a waste product. This process uses the ATP and NADPH produced by the photosynthetic reaction centres, as described above, to convert CO_2 into glycerate 3-phosphate, which can then be converted into glucose. This carbon-fixation reaction is carried out by the enzyme RuBisCO as part of the Calvin – Benson cycle. Three types of photosynthesis occur in plants, C3 carbon fixation, C4 carbon fixation and CAM photosynthesis. These differ by the route that carbon dioxide takes to the Calvin cycle, with C3 plants fixing CO_2 directly, while C4 and CAM photosynthesis incorporate the CO_2 into other compounds first, as adaptations to deal with intense sunlight and dry conditions.

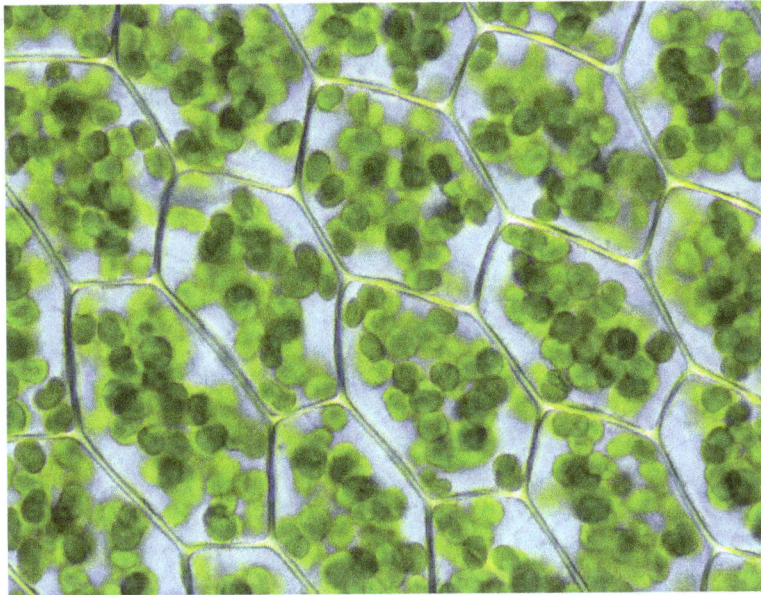

Plant cells (bounded by purple walls) filled with chloroplasts (green), which are the site of photosynthesis

In photosynthetic prokaryotes the mechanisms of carbon fixation are more diverse. Here, carbon dioxide can be fixed by the Calvin – Benson cycle, a reversed citric acid cycle, or the carboxylation of acetyl-CoA. Prokaryotic chemoautotrophs also fix CO_2 through the Calvin – Benson cycle, but use energy from inorganic compounds to drive the reaction.

Carbohydrates and Glycans

In carbohydrate anabolism, simple organic acids can be converted into monosaccharides such as glucose and then used to assemble polysaccharides such as starch. The generation of glucose from compounds like pyruvate, lactate, glycerol, glycerate 3-phosphate and amino acids is called gluconeogenesis. Gluconeogenesis converts pyruvate to glucose-6-phosphate through a series of intermediates, many of which are shared with glycolysis. However, this pathway is not simply glycolysis run in reverse, as several steps are catalyzed by non-glycolytic enzymes. This is important as it allows the formation and breakdown of glucose to be regulated separately, and prevents both pathways from running simultaneously in a futile cycle.

Although fat is a common way of storing energy, in vertebrates such as humans the fatty acids in these stores cannot be converted to glucose through gluconeogenesis as these organisms cannot convert acetyl-CoA into pyruvate; plants do, but animals do not, have the necessary enzymatic machinery. As a result, after long-term starvation, vertebrates need to produce ketone bodies from fatty acids to replace glucose in tissues such as the brain that cannot metabolize fatty acids. In other organisms such as plants and bacteria, this metabolic problem is solved using the glyoxylate cycle, which bypasses the decarboxylation step in the citric acid cycle and allows the transformation of acetyl-CoA to oxaloacetate, where it can be used for the production of glucose.

Polysaccharides and glycans are made by the sequential addition of monosaccharides by glycosyltransferase from a reactive sugar-phosphate donor such as uridine diphosphate glucose (UDP-glucose) to an acceptor hydroxyl group on the growing polysaccharide. As any of the hydroxyl groups on the ring of the substrate can be acceptors, the polysaccharides produced can have straight or branched structures. The polysaccharides produced can have structural or metabolic functions themselves, or be transferred to lipids and proteins by enzymes called oligosaccharyl-transferases.

Fatty Acids, Isoprenoids and Steroidst

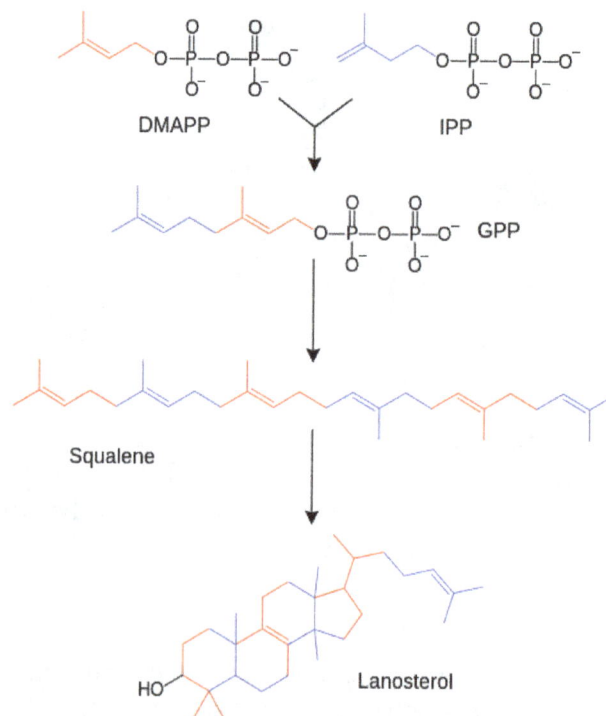

Simplified version of the steroid synthesis pathway with the intermediates isopentenyl pyrophosphate (IPP), dimethylallyl pyrophosphate (DMAPP), geranyl pyrophosphate (GPP) and squalene shown. Some intermediates are omitted for clarity.

Fatty acids are made by fatty acid synthases that polymerize and then reduce acetyl-CoA units. The acyl chains in the fatty acids are extended by a cycle of reactions that add the acyl group, reduce it to an alcohol, dehydrate it to an alkene group and then reduce it again to an alkane group. The enzymes of fatty acid biosynthesis are divided into two groups: in animals and fungi, all these fatty acid synthase reactions are carried out by a single multifunctional type I protein, while in plant plastids and bacteria separate type II enzymes perform each step in the pathway.

Terpenes and isoprenoids are a large class of lipids that include the carotenoids and form the largest class of plant natural products. These compounds are made by the assembly and modification of isoprene units donated from the reactive precursors isopentenyl pyrophosphate and dimethylallyl pyrophosphate. These precursors can be made in different ways. In animals and archaea, the mevalonate pathway produces these compounds from acetyl-CoA, while in plants and bacteria the non-mevalonate pathway uses pyruvate and glyceraldehyde 3-phosphate as substrates. One important reaction that uses these activated isoprene donors is steroid biosynthesis. Here, the isoprene units are joined together to make squalene and then folded up and formed into a set of rings to make lanosterol. Lanosterol can then be converted into other steroids such as cholesterol and ergosterol.

Proteins

Organisms vary in their ability to synthesize the 20 common amino acids. Most bacteria and plants can synthesize all twenty, but mammals can only synthesize eleven nonessential amino acids, so nine essential amino acids must be obtained from food. Some simple parasites, such as the bacteria Mycoplasma pneumoniae, lack all amino acid synthesis and take their amino acids directly from their hosts. All amino acids are synthesized from intermediates in glycolysis, the citric acid cycle, or the pentose phosphate pathway. Nitrogen is provided by glutamate and glutamine. Amino acid synthesis depends on the formation of the appropriate alpha-keto acid, which is then transaminated to form an amino acid.

Amino acids are made into proteins by being joined together in a chain of peptide bonds. Each different protein has a unique sequence of amino acid residues: this is its primary structure. Just as the letters of the alphabet can be combined to form an almost endless variety of words, amino acids can be linked in varying sequences to form a huge variety of proteins. Proteins are made from amino acids that have been activated by attachment to a transfer RNA molecule through an ester bond. This aminoacyl-tRNA precursor is produced in an ATP-dependent reaction carried out by an aminoacyl tRNA synthetase. This aminoacyl-tRNA is then a substrate for the ribosome, which joins the amino acid onto the elongating protein chain, using the sequence information in a messenger RNA.

Nucleotide Synthesis and Salvage

Nucleotides are made from amino acids, carbon dioxide and formic acid in pathways that require large amounts of metabolic energy. Consequently, most organisms have efficient systems to salvage preformed nucleotides. Purines are synthesized as nucleosides (bases attached to ribose). Both adenine and guanine are made from the precursor nucleoside inosine monophosphate, which is synthesized using atoms from the amino acids glycine, glutamine, and aspartic acid, as well as formate transferred from the coenzyme tetrahydrofolate. Pyrimidines, on the other hand, are synthesized from the base orotate, which is formed from glutamine and aspartate.

Xenobiotics and Redox Metabolism

All organisms are constantly exposed to compounds that they cannot use as foods and would be harmful if they accumulated in cells, as they have no metabolic function. These potentially damaging compounds are called xenobiotics. Xenobiotics such as synthetic drugs, natural poisons and antibiotics are detoxified by a set of xenobiotic-metabolizing enzymes. In humans, these include cytochrome P450 oxidases, UDP-glucuronosyltransferases, and glutathione S-transferases. This system of enzymes acts in three stages to firstly oxidize the xenobiotic (phase I) and then conjugate water-soluble groups onto the molecule (phase II). The modified water-soluble xenobiotic can then be pumped out of cells and in multicellular organisms may be further metabolized before being excreted (phase III). In ecology, these reactions are particularly important in microbial biodegradation of pollutants and the bioremediation of contaminated land and oil spills. Many of these microbial reactions are shared with multicellular organisms, but due to the incredible diversity of types of microbes these organisms are able to deal with a far wider range of xenobiotics than multicellular organisms, and can degrade even persistent organic pollutants such as organochloride compounds.

A related problem for aerobic organisms is oxidative stress. Here, processes including oxidative phosphorylation and the formation of disulfide bonds during protein folding produce reactive oxygen species such as hydrogen peroxide. These damaging oxidants are removed by antioxidant metabolites such as glutathione and enzymes such as catalases and peroxidases.

Thermodynamics of Living Organisms

Living organisms must obey the laws of thermodynamics, which describe the transfer of heat and work. The second law of thermodynamics states that in any closed system, the amount of entropy (disorder) cannot decrease. Although living organisms' amazing complexity appears to contradict this law, life is possible as all organisms are open systems that exchange matter and energy with their surroundings. Thus living systems are not in equilibrium, but instead are dissipative systems that maintain their state of high complexity by causing a larger increase in the entropy of their environments. The metabolism of a cell achieves this by coupling the spontaneous processes of catabolism to the non-spontaneous processes of anabolism. In thermodynamic terms, metabolism maintains order by creating disorder.

Regulation and Control

As the environments of most organisms are constantly changing, the reactions of metabolism must be finely regulated to maintain a constant set of conditions within cells, a condition called homeostasis. Metabolic regulation also allows organisms to respond to signals and interact actively with their environments. Two closely linked concepts are important for understanding how metabolic pathways are controlled. Firstly, the regulation of an enzyme in a pathway is how its activity is increased and decreased in response to signals. Secondly, the control exerted by this enzyme is the effect that these changes in its activity have on the overall rate of the pathway (the flux through the pathway). For example, an enzyme may show large changes in activity (i.e. it is highly regulated) but if these changes have little effect on the flux of a metabolic pathway, then this enzyme is not involved in the control of the pathway.

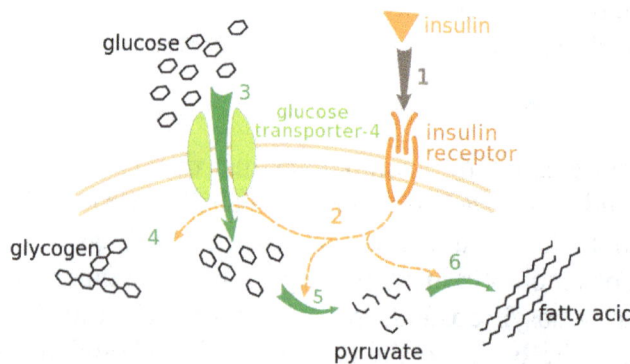

Effect of insulin on glucose uptake and metabolism. Insulin binds to its receptor (1), which in turn starts many protein activation cascades (2). These include: translocation of Glut-4 transporter to the plasma membrane and influx of glucose (3), glycogen synthesis (4), glycolysis (5) and fatty acid synthesis (6).

There are multiple levels of metabolic regulation. In intrinsic regulation, the metabolic pathway self-regulates to respond to changes in the levels of substrates or products; for example, a decrease in the amount of product can increase the flux through the pathway to compensate. This type of regulation often involves allosteric regulation of the activities of multiple enzymes in the pathway. Extrinsic control involves a cell in a multicellular organism changing its metabolism in response to signals from other cells. These signals are usually in the form of soluble messengers such as hormones and growth factors and are detected by specific receptors on the cell surface. These signals are then transmitted inside the cell by second messenger systems that often involved the phosphorylation of proteins.

A very well understood example of extrinsic control is the regulation of glucose metabolism by the hormone insulin. Insulin is produced in response to rises in blood glucose levels. Binding of the hormone to insulin receptors on cells then activates a cascade of protein kinases that cause the cells to take up glucose and convert it into storage molecules such as fatty acids and glycogen. The metabolism of glycogen is controlled by activity of phosphorylase, the enzyme that breaks down glycogen, and glycogen synthase, the enzyme that makes it. These enzymes are regulated in a reciprocal fashion, with phosphorylation inhibiting glycogen synthase, but activating phosphorylase. Insulin causes glycogen synthesis by activating protein phosphatases and producing a decrease in the phosphorylation of these enzymes.

Evolution

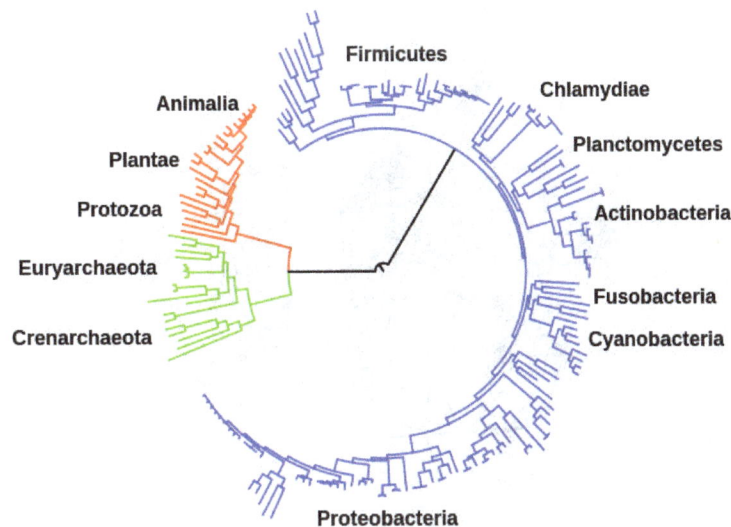

Evolutionary tree showing the common ancestry of organisms from all three domains of life. Bacteria are colored blue, eukaryotes red, and archaea green. Relative positions of some of the phyla included are shown around the tree.

The central pathways of metabolism described above, such as glycolysis and the citric acid cycle, are present in all three domains of living things and were present in the last universal ancestor. This universal ancestral cell was prokaryotic and probably a methanogen that had extensive amino acid, nucleotide, carbohydrate and lipid metabolism. The retention of these ancient pathways during later evolution may be the result of these reactions having been an optimal solution to their particular metabolic problems, with pathways such as glycolysis and the citric acid cycle producing their end products highly efficiently and in a minimal number of steps. The first pathways of enzyme-based metabolism may have been parts of purine nucleotide metabolism, while previous metabolic pathways were a part of the ancient RNA world.

Many models have been proposed to describe the mechanisms by which novel metabolic pathways evolve. These include the sequential addition of novel enzymes to a short ancestral pathway, the duplication and then divergence of entire pathways as well as the recruitment of pre-existing enzymes and their assembly into a novel reaction pathway. The relative importance of these mechanisms is unclear, but genomic studies have shown that enzymes in a pathway are likely to have a shared ancestry, suggesting that many pathways have evolved in a step-by-step fashion with novel functions created from pre-existing steps in the pathway. An alternative model comes from studies that trace the evolution of proteins' structures in metabolic networks, this has suggested that enzymes are pervasively recruited, borrowing enzymes to perform similar functions in different metabolic pathways (evident in the MANET database) These recruitment processes result in an evolutionary enzymatic mosaic. A third possibility is that some parts of metabolism might exist as "modules" that can be reused in different pathways and perform similar functions on different molecules.

As well as the evolution of new metabolic pathways, evolution can also cause the loss of metabolic functions. For example, in some parasites metabolic processes that are not essential for survival are lost and preformed amino acids, nucleotides and carbohydrates may instead be scavenged from the host. Similar reduced metabolic capabilities are seen in endosymbiotic organisms.

Investigation and Manipulation

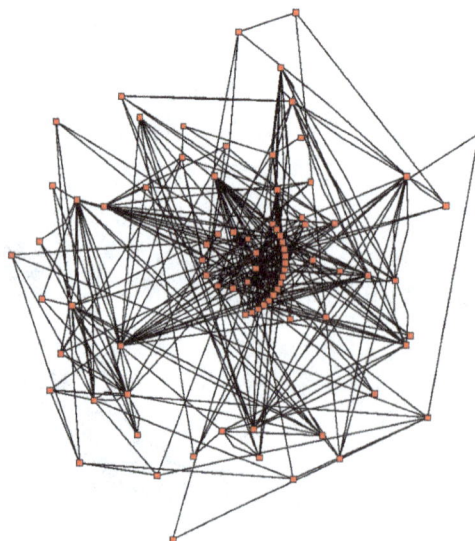

Metabolic network of the Arabidopsis thaliana citric acid cycle. Enzymes and metabolites are shown as red squares and the interactions between them as black lines.

Classically, metabolism is studied by a reductionist approach that focuses on a single metabolic pathway. Particularly valuable is the use of radioactive tracers at the whole-organism, tissue and cellular levels, which define the paths from precursors to final products by identifying radioactively labelled intermediates and products. The enzymes that catalyze these chemical reactions can then be purified and their kinetics and responses to inhibitors investigated. A parallel approach is to identify the small molecules in a cell or tissue; the complete set of these molecules is called the metabolome. Overall, these studies give a good view of the structure and function of simple metabolic pathways, but are inadequate when applied to more complex systems such as the metabolism of a complete cell.

An idea of the complexity of the metabolic networks in cells that contain thousands of different enzymes is given by the figure showing the interactions between just 43 proteins and 40 metabolites to the right: the sequences of genomes provide lists containing anything up to 45,000 genes. However, it is now possible to use this genomic data to reconstruct complete networks of biochemical reactions and produce more holistic mathematical models that may explain and predict their behavior. These models are especially powerful when used to integrate the pathway and metabolite data obtained through classical methods with data on gene expression from proteomic and DNA microarray studies. Using these techniques, a model of human metabolism has now been produced, which will guide future drug discovery and biochemical research. These models are now used in network analysis, to classify human diseases into groups that share common proteins or metabolites.

Bacterial metabolic networks are a striking example of bow-tie organization, an architecture able to input a wide range of nutrients and produce a large variety of products and complex macromolecules using a relatively few intermediate common currencies.

A major technological application of this information is metabolic engineering. Here, organisms such as yeast, plants or bacteria are genetically modified to make them more useful in biotechnology and aid the production of drugs such as antibiotics or industrial chemicals such as 1,3-propanediol and shikimic acid. These genetic modifications usually aim to reduce the amount of energy used to produce the product, increase yields and reduce the production of wastes.

History

Santorio Santorio in his steelyard balance, from Ars de statica medicina, first published 1614

The term metabolism is derived from – "Metabolismos" for "change", or "overthrow". The first documented references of metabolism were made by Ibn al-Nafis in his 1260 AD work titled Al-Risalah al-Kamiliyyah fil Siera al-Nabawiyyah (The Treatise of Kamil on the Prophet's Biography) which included the following phrase "Both the body and its parts are in a continuous state of dissolution and nourishment, so they are inevitably undergoing permanent change." The history of the scientific study of metabolism spans several centuries and has moved from examining whole animals in early studies, to examining individual metabolic reactions in modern biochemistry. The first controlled experiments in human metabolism were published by Santorio Santorio in 1614 in his book Ars de statica medicina. He described how he weighed himself before and after eating, sleep, working, sex, fasting, drinking, and excreting. He found that most of the food he took in was lost through what he called "insensible perspiration".

In these early studies, the mechanisms of these metabolic processes had not been identified and a vital force was thought to animate living tissue. In the 19th century, when studying the fermentation of sugar to alcohol by yeast, Louis Pasteur concluded that fermentation was catalyzed by substances within the yeast cells he called "ferments". He wrote that "alcoholic fermentation is an act correlated with the life and organization of the yeast cells, not with the death or putrefaction of the cells." This discovery, along with the publication by Friedrich Wöhler in 1828 of a paper on the chemical synthesis of urea, and is notable for being the first organic compound prepared from wholly inorganic precursors. This proved that the organic compounds and chemical reactions found in cells were no different in principle than any other part of chemistry.

It was the discovery of enzymes at the beginning of the 20th century by Eduard Buchner that separated the study of the chemical reactions of metabolism from the biological study of cells, and marked the beginnings of biochemistry. The mass of biochemical knowledge grew rapidly throughout the early 20th century. One of the most prolific of these modern biochemists was Hans Krebs who made huge contributions to the study of metabolism. He discovered the urea cycle and later, working with Hans Kornberg, the citric acid cycle and the glyoxylate cycle. Modern biochemical research has been greatly aided by the development of new techniques such as chromatography, X-ray diffraction, NMR spectroscopy, radioisotopic labelling, electron microscopy and molecular dynamics simulations. These techniques have allowed the discovery and detailed analysis of the many molecules and metabolic pathways in cells.

References

- Crick F (1988). "Chapter 8: The genetic code". What mad pursuit: a personal view of scientific discovery. New York: Basic Books. pp. 89–101. ISBN 0-465-09138-5.

- Mulligan PK, King RC, Stansfield WD (2006). A dictionary of genetics. Oxford [Oxfordshire]: Oxford University Press. p. 608. ISBN 0-19-530761-5.

- Griffiths AJ, Miller JH, Suzuki DT, Lewontin RC, et al., eds. (2000). "Spontaneous mutations". An Introduction to Genetic Analysis (7th ed.). New York: W. H. Freeman. ISBN 0-7167-3520-2.

- Lewis R (2005). Human Genetics: Concepts and Applications (6th ed.). Boston, Mass: McGraw Hill. pp. 227–228. ISBN 0-07-111156-5.

- Watson JD, Baker TA, Bell SP, Gann A, Levine M, Oosick R (2008). Molecular Biology of the Gene. San Francisco: Pearson/Benjamin Cummings. ISBN 0-8053-9592-X.

- Yang et al. (1990) in Michel-Beyerle, M. E., ed. Reaction centers of photosynthetic bacteria: Feldafing-II-Meeting 6. Berlin: Springer-Verlag. pp. 209–18. ISBN 3-540-53420-2.

- Yarus M (2010). Life from an RNA World: The Ancestor Within. Cambridge: Harvard University Press. p. 163. ISBN 0-674-05075-4.

- Simon M (2005). Emergent computation: emphasizing bioinformatics. New York: AIP Press/Springer Science+Business Media. pp. 105–106. ISBN 0-387-22046-1.

- Sengupta S, Higgs PG (2015). "Pathways of genetic code evolution in ancient and modern organisms". Journal of Molecular Evolution. 80: 229–243. doi:10.1007/s00239-015-9686-8.

- Taylor DJ, Ballinger MJ, Bowman SM, Bruenn JA (2013). "Virus-host co-evolution under a modified nuclear genetic code". PeerJ. 1: e50. doi:10.7717/peerj.50. PMC 3628385. PMID 23638388.

- Erives A (Aug 2011). "A model of proto-anti-codon RNA enzymes requiring L-amino acid homochirality". Journal of Molecular Evolution. 73 (1–2): 10–22. doi:10.1007/s00239-011-9453-4. PMC 3223571. PMID 21779963.

- Elzanowski A, Ostell J (2008-04-07). "The Genetic Codes". National Center for Biotechnology Information (NCBI). Retrieved 2010-03-10.

Allied Fields of Biochemistry

This chapter will discuss all the allied fields of biochemistry. The major topics discussed here are molecular biology, cell biology, biotechnology and bioluminescence. All the topics mentioned in the chapter are rapidly growing branches of biochemistry and are, therefore, an important part of this field.

Molecular Biology

Molecular biology concerns the molecular basis of biological activity between biomolecules in the various systems of a cell, including the interactions between DNA, RNA and proteins and their biosynthesis, as well as the regulation of these interactions. Writing in Nature in 1961, William Astbury described molecular biology as:

"...not so much a technique as an approach, an approach from the viewpoint of the so-called basic sciences with the leading idea of searching below the large-scale manifestations of classical biology for the corresponding molecular plan. It is concerned particularly with the forms of biological molecules and [...] is predominantly three-dimensional and structural—which does not mean, however, that it is merely a refinement of morphology. It must at the same time inquire into genesis and function."

Relationship to Other Biological Sciences

Researchers in molecular biology use specific techniques native to molecular biology but increasingly combine these with techniques and ideas from genetics and biochemistry. There is not a defined line between these disciplines. The figure to the right is a schematic that depicts one possible view of the relationship between the fields:

- Biochemistry is the study of the chemical substances and vital processes occurring in live organisms. Biochemists focus heavily on the role, function, and structure of biomolecules. The study of the chemistry behind biological processes and the synthesis of biologically active molecules are examples of biochemistry.

- Genetics is the study of the effect of genetic differences on organisms. This can often be inferred by the absence of a normal component (e.g. one gene). The study of "mutants" – organisms which lack one or more functional components with respect to the so-called "wild type" or normal phenotype. Genetic interactions (epistasis) can often confound simple interpretations of such "knockout" studies.

- Molecular biology is the study of molecular underpinnings of the processes of replication, transcription, translation, and cell function. The central dogma of molecular biology where genetic material is transcribed into RNA and then translated into protein, despite being an oversimplified picture of molecular biology, still provides a good starting point for understanding the field. This picture, however, is undergoing revision in light of emerging novel roles for RNA.

Much of the work in molecular biology is quantitative, and recently much work has been done at the interface of molecular biology and computer science in bioinformatics and computational biology. As of the early 2000s, the study of gene structure and function, molecular genetics, has been among the most prominent sub-field of molecular biology.Increasingly many other loops of biology focus on molecules, either directly studying their interactions in their own right such as in cell biology and developmental biology, or indirectly, where the techniques of molecular biology are used to infer historical attributes of populations or species, as in fields in evolutionary biology such as population genetics and phylogenetics. There is also a long tradition of studying biomolecules "from the ground up" in biophysics.

Techniques of Molecular Biology

For more extensive list on protein methods. For more extensive list on nucleic acid methods.

Since the late 1950s and early 1960s, molecular biologists have learned to characterize, isolate, and manipulate the molecular components of cells and organisms. These components include DNA, the repository of genetic information; RNA, a close relative of DNA whose functions range from serving as a temporary working copy of DNA to actual structural and enzymatic functions as well as a functional and structural part of the translational apparatus, the ribosome; and proteins, the major structural and enzymatic type of molecule in cells.

Molecular Cloning

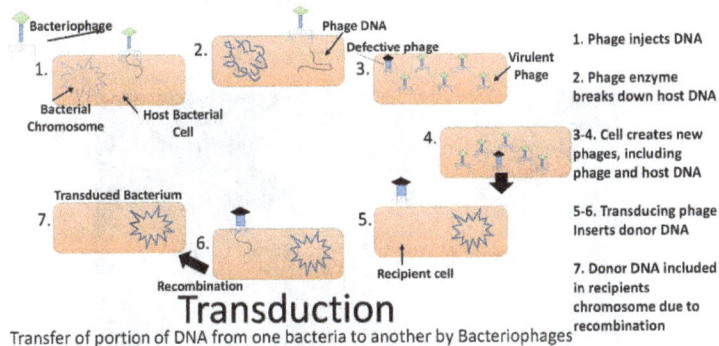

Transduction image

One of the most basic techniques of molecular biology to study protein function is molecular cloning. In this technique, DNA coding for a protein of interest is cloned (using PCR and/or restriction enzymes) into a plasmid (known as an expression vector). A vector has 3 distinctive features: an origin of replication, a multiple cloning site (MCS), and a selective marker (usually antibiotic resistance). The origin of replication will have promoter regions upstream from the replication/transcription start site.

This plasmid can be inserted into either bacterial or animal cells. Introducing DNA into bacterial cells can be done by transformation (via uptake of naked DNA), conjugation (via cell-cell contact) or by transduction (via viral vector). Introducing DNA into eukaryotic cells, such as animal cells, by physical or chemical means is called transfection. Several different transfection techniques are available, such as calcium phosphate transfection, electroporation, microinjection and liposome transfection. DNA can also be introduced into eukaryotic cells using viruses or bacteria as carriers, the latter is sometimes called bactofection and in particular uses Agrobacterium tumefaciens. The plasmid may be integrated into the genome, resulting in a stable transfection, or may remain independent of the genome, called transient transfection.

In either case, DNA coding for a protein of interest is now inside a cell, and the protein can now be expressed. A variety of systems, such as inducible promoters and specific cell-signaling factors, are available to help express the protein of interest at high levels. Large quantities of a protein can then be extracted from the bacterial or eukaryotic cell. The protein can be tested for enzymatic activity under a variety of situations, the protein may be crystallized so its tertiary structure can be studied, or, in the pharmaceutical industry, the activity of new drugs against the protein can be studied.

Polymerase Chain Reaction (PCR)

Polymerase chain reaction is an extremely versatile technique for copying DNA. In brief, PCR allows a specific DNA sequence to be copied or modified in predetermined ways. The reaction is extremely powerful and under perfect conditions could amplify 1 DNA molecule to become 1.07 Billion molecules in less than 2 hours. The PCR technique can be used to introduce restriction enzyme sites to ends of DNA molecules, or to mutate (change) particular bases of DNA, the latter is a method referred to as site-directed mutagenesis. PCR can also be used to determine whether a particular DNA fragment is found in a cDNA library. PCR has many variations, like reverse transcription PCR (RT-PCR) for amplification of RNA, and, more recently, quantitative PCR which allow for quantitative measurement of DNA or RNA molecules.

Gel Electrophoresis

Two percent Agarose Gel in Borate Buffer cast in a Gel Tray (Front, angled)

Gel electrophoresis is one of the principal tools of molecular biology. The basic principle is that DNA, RNA, and proteins can all be separated by means of an electric field and size. In agarose gel electrophoresis, DNA and RNA can be separated on the basis of size by running the DNA through an electrically charged agarose gel. Proteins can be separated on the basis of size by using an SDS-PAGE gel, or on the basis of size and their electric charge by using what is known as a 2D gel electrophoresis.

Macromolecule Blotting and Probing

The terms northern, western and eastern blotting are derived from what initially was a molecular biology joke that played on the term Southern blotting, after the technique described by Edwin Southern for the hybridisation of blotted DNA. Patricia Thomas, developer of the RNA blot which then became known as the northern blot, actually didn't use the term. Further combinations of these techniques produced such terms as southwesterns (protein-DNA hybridizations), northwesterns (to detect protein-RNA interactions) and farwesterns (protein-protein interactions), all of which are presently found in the literature.

Southern Blotting

Named after its inventor, biologist Edwin Southern, the Southern blot is a method for probing for the presence of a specific DNA sequence within a DNA sample. DNA samples before or after restriction enzyme (restriction endonuclease) digestion are separated by gel electrophoresis and then transferred to a membrane by blotting via capillary action. The membrane is then exposed to a labeled DNA probe that has a complement base sequence to the sequence on the DNA of interest. Most original protocols used radioactive labels, however non-radioactive alternatives are now available. Southern blotting is less commonly used in laboratory science due to the capacity of other techniques, such as PCR, to detect specific DNA sequences from DNA samples. These blots are still used for some applications, however, such as measuring transgene copy number in transgenic mice, or in the engineering of gene knockout embryonic stem cell lines.

Northern Blotting

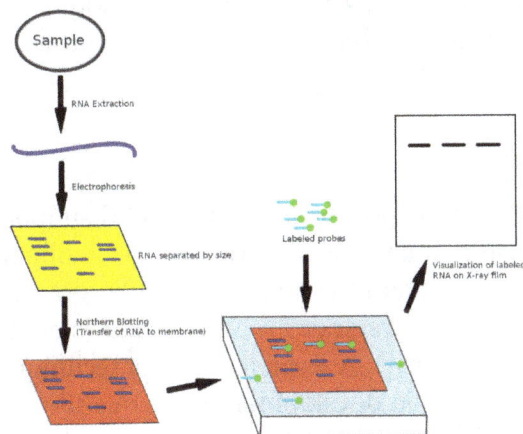

Northern blot diagram

The northern blot is used to study the expression patterns of a specific type of RNA molecule as relative comparison among a set of different samples of RNA. It is essentially a combination of denaturing RNA gel electrophoresis, and a blot. In this process RNA is separated based on size and is then transferred to a membrane that is then probed with a labeled complement of a sequence of interest. The results may be visualized through a variety of ways depending on the label used; however, most result in the revelation of bands representing the sizes of the RNA detected in sample. The intensity of these bands is related to the amount of the target RNA in the samples analyzed. The procedure is commonly used to study when and how much gene expression is occurring by measuring how much of that RNA is present in different samples. It is one of the most basic tools for determining at what time, and under what conditions, certain genes are expressed in living tissues.

Western Blotting

Antibodies to most proteins can be created by injecting small amounts of the protein into an animal such as a mouse, rabbit, sheep, or donkey (polyclonal antibodies) or produced in cell culture (monoclonal antibodies). These antibodies can be used for a variety of analytical and preparative techniques.

In western blotting, proteins are first separated by size, in a thin gel sandwiched between two glass plates in a technique known as SDS-PAGE (sodium dodecyl sulfate polyacrylamide gel electrophoresis). The proteins in the gel are then transferred to a polyvinylidene fluoride (PVDF), nitrocellulose, nylon, or other support membrane. This membrane can then be probed with solutions of antibodies. Antibodies that specifically bind to the protein of interest can then be visualized by a variety of techniques, including colored products, chemiluminescence, or autoradiography. Often, the antibodies are labeled with enzymes. When a chemiluminescent substrate is exposed to the enzyme it allows detection. Using western blotting techniques allows not only detection but also quantitative analysis. Analogous methods to western blotting can be used to directly stain specific proteins in live cells or tissue sections. However, these immunostaining methods, such as FISH, are used more often in cell biology research.

Eastern Blotting

The Eastern blotting technique is used to detect post-translational modification of proteins. Proteins blotted on to the PVDF or nitrocellulose membrane are probed for modifications using specific substrates.

Microarrays

A DNA microarray is a collection of spots attached to a solid support such as a microscope slide where each spot contains one or more single-stranded DNA oligonucleotide fragment. Arrays make it possible to put down large quantities of very small (100 micrometre diameter) spots on a single slide. Each spot has a DNA fragment molecule that is complementary to a single DNA sequence (similar to Southern blotting). A variation of this technique allows the gene expression of an organism at a particular stage in development to be qualified (expression profiling). In this technique the RNA in a tissue is isolated and converted to labeled cDNA. This cDNA is then hybridized to the fragments on the array and visualization of the hybridization can be done. Since multiple arrays can be made with exactly the same position of fragments they are particularly useful for comparing the gene expression of two differ rent tissues, such as a healthy and cancerous tissue. Also, one can measure what genes are expressed and how that expression changes with time or with other factors. For instance, the common baker's yeast, Saccharomyces cerevisiae, contains about 7000 genes; with a microarray, one can measure qualitatively how each gene is expressed, and how that expression changes, for example, with a change in temperature. There are many different ways to fabricate microarrays; the most common are silicon chips, microscope slides with spots of ~ 100 micrometre diameter, custom arrays, and arrays with larger spots on porous membranes (macroarrays). There can be anywhere from 100 spots to more than 10,000 on a given array.Arrays can also be made with molecules other than DNA. For example, an antibody array can be used to determine what proteins or bacteria are present in a blood sample.

Allele-Specific Oligonucleotide

Allele-specific oligonucleotide (ASO) is a technique that allows detection of single base mutations without the need for PCR or gel electrophoresis. Short (20-25 nucleotides in length), labeled probes are exposed to the non-fragmented target DNA. Hybridization occurs with high specificity due to the short length of the probes and even a single base change will hinder hybridization. The target DNA is then washed and the labeled probes that didn't hybridize are removed. The target DNA is then analyzed for the presence of the probe via radioactivity or fluorescence. In this experiment, as in most molecular biology techniques, a control must be used to ensure successful experimentation. The Illumina Methylation Assay is an example of a method that takes advantage of the ASO technique to measure one base pair differences in sequence.

Antiquated Technologies

In molecular biology, procedures and technologies are continually being developed and older technologies abandoned. For example, before the advent of DNA gel electrophoresis (agarose or polyacrylamide), the size of DNA molecules was typically determined by rate sedimentation in sucrose gradients, a slow and labor-intensive technique requiring expensive instrumentation; prior to sucrose gradients, viscometry was used.Aside from their historical interest, it is often worth knowing about older technology, as it is occasionally useful to solve another new problem for which the newer technique is inappropriate.

History

While molecular biology was established in the 1930s, the term was coined by Warren Weaver in 1938. Weaver was the director of Natural Sciences for the Rockefeller Foundation at the time and believed that biology was about to undergo a period of significant change given recent advances in fields such as X-ray crystallography. He therefore channeled significant amounts of (Rockefeller Institute) money into biological fields.

Clinical Significance

Clinical research and medical therapies arising from molecular biology are partly covered under gene therapy. The use of molecular biology or molecular cell biology approaches in medicine is now called molecular medicine. Molecular biology also plays important role in understanding formations, actions, regulations of various parts of cells which can be used efficiently for targeting new drugs, diagnosis of disease, physiology of the Cell.

Cell Biology

Cell biology (formerly called cytology) and otherwise known as molecular biology, is a branch of biology that studies the different structures and functions of the cell and focuses mainly on the idea of the cell as the basic unit of life. Cell biology explains the structure, organization of the organelles they contain, their physiological properties, metabolic processes, signaling pathways, life cycle, and interactions with their environment. This is done both on a microscopic and molecular level as it encompasses prokaryotic cells and eukaryotic cells. Knowing the components of cells and how cells work is fundamental to all biological sciences it is also essential for research in bio-medical fields such as cancer, and other diseases. Research in cell biology is closely related to genetics, bio-chemistry, molecular biology, immunology, and developmental biology.

Internal Cellular Structures

Chemical and Molecular Environment

The study of the cell is done on a molecular level; however, most of the processes within the cell is made up of a mixture of small organic molecules, inorganic ions, hormones, and water. Approximately 75-85% of the cell's volume is due to water making it an indispensable solvent as a result of its polarity and structure. These molecules within the cell, which operate as substrates, provide a suitable environment for the cell to carry out metabolic reactions and signalling.

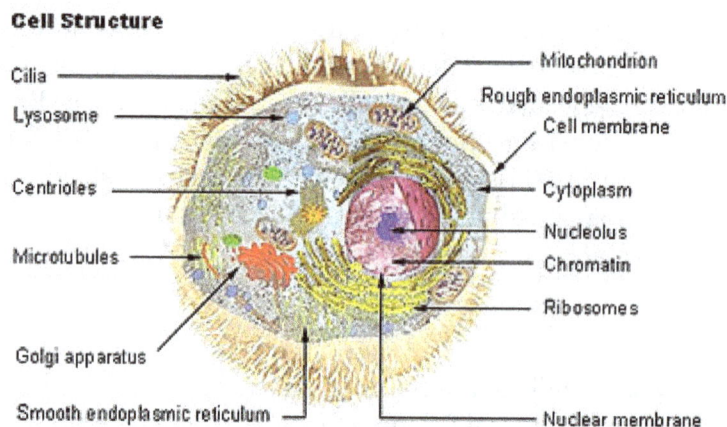

Cell Structure

Cilia	Mitochondrion
Lysosome	Rough endoplasmic reticulum
	Cell membrane
Centrioles	Cytoplasm
	Nucleolus
Microtubules	Chromatin
	Ribosomes
Golgi apparatus	
Smooth endoplasmic reticulum	Nuclear membrane

Understanding the cell in terms of its generalized structure and molecular components.

The cell shape varies among the different types of organisms, and are thus then classified into two categories: eukaryotes and prokaryotes. In the case of eukaryotic cells - which are made up of animal, plant, fungi, and protozoa cells - the shapes are generally round and spherical, while for prokaryotic cells – which are composed of bacteria and archaea - the shapes are: spherical (cocci), rods (bacillus), curved (vibrio), and spirals (spirochetes).

Cell biology focuses more on the study of eukaryotic cells, and their signalling pathways, rather than on prokaryotes which is covered under microbiology. The main constituents of the general molecular composition of the cell includes: proteins and lipids which are either free flowing or membrane bound, along with different internal compartments known as organelles. This environment of the cell is made up of hydrophilic and hydrophobic regions which allows for the exchange of the above-mentioned molecules and ions. The hydrophilic regions of the cell are mainly on the inside and outside of the cell, while the hydrophobic regions are within the phospholipid bilayer of the cell membrane. The cell membrane consists of lipids and proteins which accounts for its hydrophobicity as a result of being non-polar substances. Therefore, in order for these molecules to participate in reactions, within the cell, they need to be able to cross this membrane layer to get into the cell. They accomplish this process of gaining access to the cell via: osmotic pressure, diffusion, concentration gradients, and membrane channels. Inside of the cell are extensive internal sub-cellular membrane-bounded compartments called organelles.

Organelles

- Centrosome - an associated pair of cylindrical shaped protein structures (centrioles) that organize microtubules and aid in forming the mitotic spindle during cell division in eukaryotes

- Cell membrane (plasma membrane) - the part of the cell which separates the cells from the outside environment and protects the cell, as well as regulating what goes in and out of the cell

- Cell wall - extra layer of protection and gives structural support (only found in plant cells)

- Chloroplast - key organelle for photosynthesis (only found in plant cells)

- Cilium - motile structure of eukaryotes having a cytoskeleton, the axoneme.

- Cytoplasm - contents of the main fluid-filled space inside cells, chemical reactions also happen in this jelly-like substance.

- Cytoskeleton - protein filaments inside cells (microfilaments, microtubules, and intermediate filaments)

- Endoplasmic reticulum (rough) - major site of membrane protein synthesis

- Endoplasmic reticulum (smooth) - major site of lipid synthesis

- Endosomes - vesicles that traffic membrane and intra and extra cellular contents for recycling or degradation by lysosomes

- Flagellum - motile structure of bacteria, archaea and eukaryotes

- Golgi apparatus - site of protein glycosylation in the endomembrane system

- Lipid bilayer - fundamental organizational structure of cell membranes

- Lysosome - acidic organelle that breaks down cellular waste products and debris into simple compounds (only found in animal cells)

- Microvilli - increases surface area for absorption of nutrients from surrounding medium

- Mitochondrion - major energy-producing organelle by releasing energy in the form of ATP

- Nucleus - contains chromosomes composed of DNA, the building block of life. Nuclear Architecture is important for dictating nuclear function.

- Organelle - term used for major subcellular structures

- Peroxisomes - a very small organelle that uses oxygen to breakdown and detoxify long fatty acids and other molecules

- Pili - also called fimbria is used for conjugation and sometimes movement

- Ribosome - RNA and protein complex required for protein synthesis in cells

- Starch grain - found in the cytoplasm of a typical plant cell, it stores chemical energy of the plant.

- Vacuole - contain cell sap (only found in plant cells)

- Vesicle - small membrane-bounded spheres inside cells

cell surface membrane protects the cell

Processes

Growth and Development

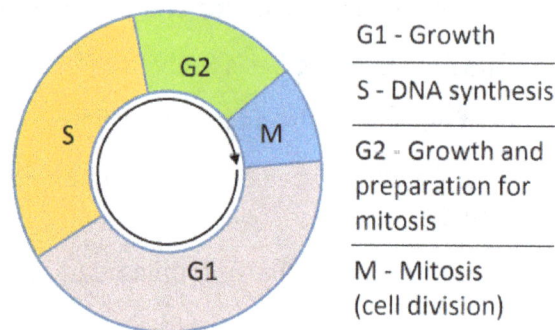

G1 - Growth	
S - DNA synthesis	
G2 - Growth and preparation for mitosis	
M - Mitosis (cell division)	

General concept of the cell cycle.

The growth process of the cell does not refer to the size of the cell, but instead the density of the number of cells present in the organism at a given time. Cell growth pertains to the increase in the number of cells present in an organism as it grows and develops; as the organism gets larger so too does the number of cells present. Cells are the foundation of all organisms, they are the fundamental unit of life. The growth and development of the cell are essential for the maintenance of the host, and survival of the organisms. For this process the cell goes through the steps of the cell cycle and development which involves cell growth, DNA replication, cell division, regeneration, specialization, and cell death. The cell cycle is divided into four distinct phases, G1, S, G2, and M. The G phases – which is the cell growth phase - makes up approximately 95% of the cycle. The proliferation of cells is instigated by progenitors, the cells then differentiate to become specialized, where specialized cells of the same type aggregate to form tissues, then organs and ultimately systems. The G phases along with the S phase – DNA replication, damage and repair - are considered to be the interphase portion of the cycle. While the M phase (mitosis and cytokinesis) is the cell division portion of the cycle. The cell cycle is regulated by a series of signalling factors and complexes such as CDK's, kinases, and p53. to name a few. When the cell has completed its growth process, and if it is found to be damaged or altered it undergoes cell death, either by apoptosis or necrosis, to eliminate the threat it cause to the organism's survival.

Other Cellular Processes

- Active transport and Passive transport - Movement of molecules into and out of cells.

- Autophagy - The process whereby cells "eat" their own internal components or microbial invaders.

- Adhesion - Holding together cells and tissues.

- Cell movement - Chemotaxis, contraction, cilia and flagella.

- Cell signaling - Regulation of cell behavior by signals from outside.

- Division - By which cells reproduce either by mitosis (to produce clones of the parent cell) or Meiosis (to produce haploid gametes)

- DNA repair - Cell death and cell senescence.

- Metabolism - Glycolysis, respiration, photosynthesis, and chemosynthesis.

- Signalling - The process by which the activities in the cell are regulated

- Transcription and mRNA splicing - Gene expression.

Techniques Used to Study Cells

Green Fluorescent Protein and Fluorescence Microscope

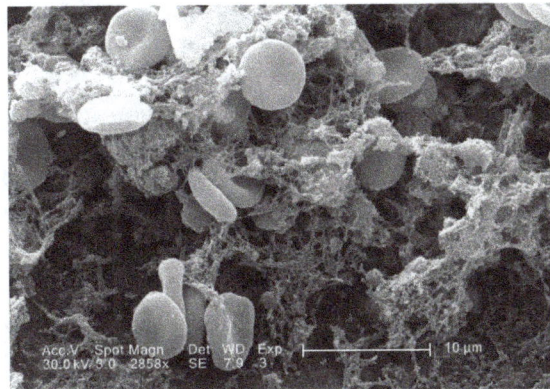

Electron micrograph of blood cells clotting.

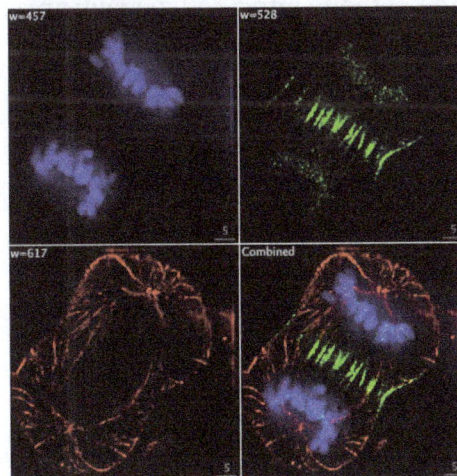

Cell division studied using fluorescence to stain specific structures

Cells may be observed under the microscope, using several different techniques; these include optical microscopy, transmission electron microscopy, scanning electron microscopy, fluorescence microscopy, and confocal microscopy.

There are several different methods used in the study of cells:

- Cell culture is the basic technique of growing cells in a laboratory independent of an organism.

- Immunostaining, also known as immunohistochemistry, is a specialized histological method used to localize proteins in cells or tissue slices. Unlike regular histology, which uses stains to identify cells, cellular components or protein classes, immunostaining requires the reaction of an antibody directed against the protein of interest within the tissue or cell. Through the use of proper controls and published protocols (need to add reference links here), specificity of the antibody-antigen reaction can be achieved. Once this complex is formed, it is identified via either a "tag" attached directly to the antibody, or added in an additional technical step. Commonly used "tags" include fluorophores or enzymes. In the case of the former, detection of the location of the "immuno-stained" protein occurs via fluorescence microscopy. With an enzymatic tag, such as horse radish peroxidase, a chemical reaction is carried out that results in a dark color in the location of the protein of interest. This darkened pattern is then detected using light microscopy.

- Computational genomics is used to find patterns in genomic information

- DNA microarrays identify changes in transcript levels between different experimental conditions.

- Gene knockdown mutates a selected gene.

- In situ hybridization shows which cells are expressing a particular RNA transcript.

- PCR can be used to determine how many copies of a gene are present in a cell.

- Transfection introduces a new gene into a cell, usually an expression construct

Purification of cells and their parts Purification may be performed using the following methods:

- Cell fractionation

  o Release of cellular organelles by disruption of cells.

  o Separation of different organelles by centrifugation.

- Flow cytometry

- Immunoprecipitation

  o The binding of an antibody to a target protein

  o Collection of the target protein through elution

- Proteins extracted from cell membranes by detergents and salts or other kinds of chemicals.

Career and Related Fields

Practical job applications for a degree in Cell Molecular Biology includes the following.

• Agriculture	• Genetic Engineering	• Research Assistant
• Bioinformatics & Medical Informatics	• Government Jobs	• Research Coordinator
• Biomedical Research	• Healthcare Industries	• Research Scientist
• Biotechnology	• Health Policies	• Research Technician
• Biotechnology Policy	• Health Law	• Science Administration
• Brewery Production	• Laboratory Technician	• Science Education
• Business	• Medical Research	• Science Journalism
• Clinical Diagnostics	• Medical School	• Science Policy
• Colleges and University Teaching	• Medical Technology	• Scientific Editing
• Crop improvement	• Microbiologist	• Scientific Illustrator
• Elementary, Middle and High School Teaching	• Nutrition	• Scientific Instrumentation Companies
• Environmental Technician/Consultant	• Patent Law	• Scientific Supply Representative
• Food Production and Safety	• Patient Care	• Teacher
• Forensic Scientist	• Pest control	• Technical writing
• Geneticist	• Pharmaceutical Science	• Toxicology
	• Pharmaceutical Sales Representative	• Training and outreach in Industry and Government
	• Physician Assistant	
	• Product Quality and Safety	

Notable Cell Biologists

• Jean Baptiste Carnoy	• Marc Kirschner	• Jan Evangelista Purkyně
• Peter Agre	• Anton van Leeuwenhoek	
• Günter Blobel	• Ira Mellman	
• Robert Brown	• Peter D. Mitchell	
• Geoffrey M. Cooper	• Rudolf Virchow	
• Christian de Duve	• Paul Nurse	
• Robert Hooke	• George Emil Palade	
• H. Robert Horvitz	• Keith R. Porter	

Czech anatomist Jan Evangelista Purkyně is best known for his 1837 discovery of Purkinje cells.
- Ray Rappaport
- Michael Swann
- Roger Tsien
- Edmund Beecher Wilson
- Kenneth R. Miller
- Matthias Jakob Schleiden
- Theodor Schwann

Biotechnology

Biotechnology is the use of living systems and organisms to develop or make products, or "any technological application that uses biological systems, living organisms or derivatives thereof, to make or modify products or processes for specific use" (UN Convention on Biological Diversity, Art. 2). Depending on the tools and applications, it often overlaps with the (related) fields of bio-engineering, biomedical engineering, biomanufacturing, etc.

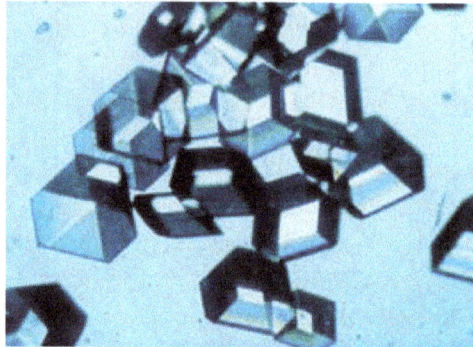

Insulin crystals

For thousands of years, humankind has used biotechnology in agriculture, food production, and medicine. The term is largely believed to have been coined in 1919 by Hungarian engineer Károly Ereky. In the late 20th and early 21st century, biotechnology has expanded to include new and diverse sciences such as genomics, recombinant gene techniques, applied immunology, and development of pharmaceutical therapies and diagnostic tests.

Definitions

The wide concept of "biotech" or "biotechnology" encompasses a wide range of procedures for modifying living organisms according to human purposes, going back to domestication of animals, cultivation of the plants, and "improvements" to these through breeding programs that employ artificial selection and hybridization. Modern usage also includes genetic engineering as well as cell and tissue culture technologies. The American Chemical Society defines biotechnology as the application of biological organisms, systems, or processes by various industries to learning about the science of life and the improvement of the value of materials and organisms such as pharmaceuticals, crops, and livestock. As per European Federation of Biotechnology, Biotechnology is the integration of natural science and organisms, cells, parts thereof, and molecular analogues for products and services. Biotechnology also writes on the pure biological sciences (animal cell culture, biochemistry, cell biology, embryology, genetics, microbiology, and molecular biology). In many instances, it is also dependent on knowledge and methods from outside the sphere of biology including:

- bioinformatics, a new brand of computer science
- bioprocess engineering
- biorobotics
- chemical engineering

Conversely, modern biological sciences (including even concepts such as molecular ecology) are intimately entwined and heavily dependent on the methods developed through biotechnology and what is commonly thought of as the life sciences industry. Biotechnology is the research and development in the laboratory using bioinformatics for exploration, extraction, exploitation and production from any living organisms and any source of biomass by means of biochemical engineering where high value-added products could be planned (reproduced by biosynthesis, for example), forecasted, formulated, developed, manufactured and marketed for the purpose of sustainable operations (for the return from bottomless initial investment on R & D) and gaining durable patents rights (for exclusives rights for sales, and prior to this to receive national and international approval from the results on animal experiment and human experiment, especially on the pharmaceutical branch of biotechnology to prevent any undetected side-effects or safety concerns by using the products).

By contrast, bioengineering is generally thought of as a related field that more heavily emphasizes higher systems approaches (not necessarily the altering or using of biological materials directly) for interfacing with and utilizing living things. Bioengineering is the application of the principles of engineering and natural sciences to tissues, cells and molecules. This can be considered as the use of knowledge from working with and manipulating biology to achieve a result that can improve functions in plants and animals. Relatedly, biomedical engineering is an overlapping field that often draws upon and applies biotechnology (by various definitions), especially in certain subfields of biomedical and/or chemical engineering such as tissue engineering, biopharmaceutical engineering, and genetic engineering.

History

Brewing was an early application of biotechnology

Although not normally what first comes to mind, many forms of human-derived agriculture clearly fit the broad definition of "utilizing a biotechnological system to make products". Indeed, the cultivation of plants may be viewed as the earliest biotechnological enterprise.

Agriculture has been theorized to have become the dominant way of producing food since the Neolithic Revolution. Through early biotechnology, the earliest farmers selected and bred the best suited crops, having the highest yields, to produce enough food to support a growing population. As crops and fields became increasingly large and difficult to maintain, it was discovered that specific organisms and their by-products could effectively fertilize, restore nitrogen, and control pests. Throughout the history of agriculture, farmers have inadvertently altered the genetics of their crops through introducing them to new environments and breeding them with other plants — one of the first forms of biotechnology.

These processes also were included in early fermentation of beer. These processes were introduced in early Mesopotamia, Egypt, China and India, and still use the same basic biological methods. In brewing, malted grains (containing enzymes) convert starch from grains into sugar and then adding specific yeasts to produce beer. In this process, carbohydrates in the grains were broken down into alcohols such as ethanol. Later other cultures produced the process of lactic acid fermentation which allowed the fermentation and preservation of other forms of food, such as soy sauce. Fermentation was also used in this time period to produce leavened bread. Although the process of fermentation was not fully understood until Louis Pasteur's work in 1857, it is still the first use of biotechnology to convert a food source into another form.

Before the time of Charles Darwin's work and life, animal and plant scientists had already used selective breeding. Darwin added to that body of work with his scientific observations about the ability of science to change species. These accounts contributed to Darwin's theory of natural selection.

For thousands of years, humans have used selective breeding to improve production of crops and livestock to use them for food. In selective breeding, organisms with desirable characteristics are mated to produce offspring with the same characteristics. For example, this technique was used with corn to produce the largest and sweetest crops.

In the early twentieth century scientists gained a greater understanding of microbiology and explored ways of manufacturing specific products. In 1917, Chaim Weizmann first used a pure microbiological culture in an industrial process, that of manufacturing corn starch using Clostridium acetobutylicum, to produce acetone, which the United Kingdom desperately needed to manufacture explosives during World War I.

Biotechnology has also led to the development of antibiotics. In 1928, Alexander Fleming discovered the mold Penicillium. His work led to the purification of the antibiotic compound formed by the mold by Howard Florey, Ernst Boris Chain and Norman Heatley – to form what we today know as penicillin. In 1940, penicillin became available for medicinal use to treat bacterial infections in humans.

The field of modern biotechnology is generally thought of as having been born in 1971 when Paul Berg's (Stanford) experiments in gene splicing had early success. Herbert W. Boyer (Univ. Calif. at San Francisco) and Stanley N. Cohen (Stanford) significantly advanced the new technology in 1972 by transferring genetic material into a bacterium, such that the imported material would be reproduced. The commercial viability of a biotechnology industry was significantly expanded on June 16, 1980, when the United States Supreme Court ruled that a genetically modified microorganism could be patented in the case of Diamond v. Chakrabarty. Indian-born Ananda Chakrabarty, working for General Electric, had modified a bacterium (of the Pseudomonas genus) capable of breaking down crude oil, which he proposed to use in treating oil spills. (Chakrabarty's work did not involve gene manipulation but rather the transfer of entire organelles between strains of the Pseudomonas bacterium.

Revenue in the industry is expected to grow by 12.9% in 2008. Another factor influencing the biotechnology sector's success is improved intellectual property rights legislation—and enforcement—worldwide, as well as strengthened demand for medical and pharmaceutical products to cope with an ageing, and ailing, U.S. population.

Rising demand for biofuels is expected to be good news for the biotechnology sector, with the Department of Energy estimating ethanol usage could reduce U.S. petroleum-derived fuel consumption by up to 30% by 2030. The biotechnology sector has allowed the U.S. farming industry to rapidly increase its supply of corn and soybeans—the main inputs into biofuels—by developing genetically modified seeds which are resistant to pests and drought. By boosting farm productivity, biotechnology plays a crucial role in ensuring that biofuel production targets are met.

Examples

A rose plant that began as cells grown in a tissue culture

Biotechnology has applications in four major industrial areas, including health care (medical), crop production and agriculture, non food (industrial) uses of crops and other products (e.g. biodegradable plastics, vegetable oil, biofuels), and environmental uses.

For example, one application of biotechnology is the directed use of organisms for the manufacture of organic products (examples include beer and milk products). Another example is using naturally present bacteria by the mining industry in bioleaching. Biotechnology is also used to recycle, treat waste, clean up sites contaminated by industrial activities (bioremediation), and also to produce biological weapons.

A series of derived terms have been coined to identify several branches of biotechnology; for example:

- Bioinformatics is an interdisciplinary field which addresses biological problems using computational techniques, and makes the rapid organization as well as analysis of biological data possible. The field may also be referred to as computational biology, and can be defined as, "conceptualizing biology in terms of molecules and then applying informatics techniques to understand and organize the information associated with these molecules, on a large scale." Bioinformatics plays a key role in various areas, such as functional genomics, structural genomics, and proteomics, and forms a key component in the biotechnology and pharmaceutical sector.

- Blue biotechnology is a term that has been used to describe the marine and aquatic applications of biotechnology, but its use is relatively rare.

- Green biotechnology is biotechnology applied to agricultural processes. An example would be the selection and domestication of plants via micropropagation. Another example is the designing of transgenic plants to grow under specific environments in the presence (or absence) of chemicals. One hope is that green biotechnology might produce more environmentally friendly solutions than traditional industrial agriculture. An example of this is the engineering of a plant to express a pesticide, thereby ending the need of external application of pesticides. An example of this would be Bt corn. Whether or not green biotechnology products such as this are ultimately more environmentally friendly is a topic of considerable debate.

- Red biotechnology is applied to medical processes. Some examples are the designing of organisms to produce antibiotics, and the engineering of genetic cures through genetic manipulation.

- White biotechnology, also known as industrial biotechnology, is biotechnology applied to industrial processes. An example is the designing of an organism to produce a useful chemical. Another example is the using of enzymes as industrial catalysts to either produce valuable chemicals or destroy hazardous/polluting chemicals. White biotechnology tends to consume less in resources than traditional processes used to produce industrial goods.

The investment and economic output of all of these types of applied biotechnologies is termed as "bioeconomy".

Medicine

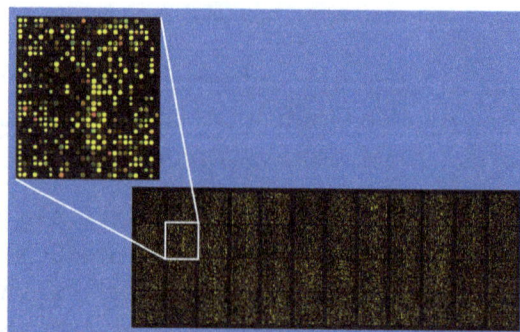

DNA microarray chip – some can do as many as a million blood tests at once

In medicine, modern biotechnology finds applications in areas such as pharmaceutical drug discovery and production, pharmacogenomics, and genetic testing (or genetic screening).

Pharmacogenomics (a combination of pharmacology and genomics) is the technology that analyses how genetic makeup affects an individual's response to drugs. It deals with the influence of genetic variation on drug response in patients by correlating gene expression or single-nucleotide polymorphisms with a drug's efficacy or toxicity. By doing so, pharmacogenomics aims to develop rational means to optimize drug therapy, with respect to the patients' genotype, to ensure maximum efficacy with minimal adverse effects. Such approaches promise the advent of "personalized medicine"; in which drugs and drug combinations are optimized for each individual's unique genetic makeup.

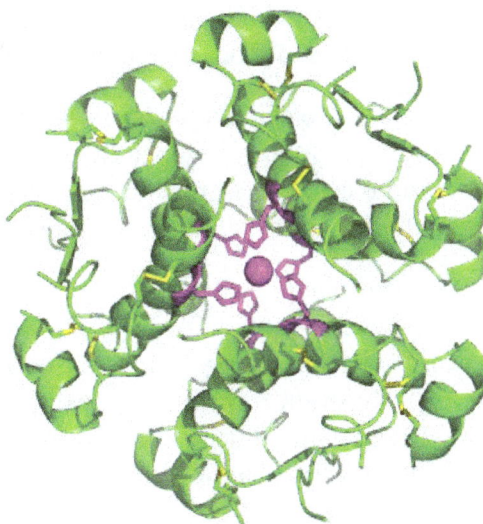

Computer-generated image of insulin hexamers highlighting the threefold symmetry, the zinc ions holding it together, and the histidine residues involved in zinc binding.

Biotechnology has contributed to the discovery and manufacturing of traditional small molecule pharmaceutical drugs as well as drugs that are the product of biotechnology – biopharmaceutics. Modern biotechnology can be used to manufacture existing medicines relatively easily and cheaply. The first genetically engineered products were medicines designed to treat human diseases. To cite one example, in 1978 Genentech developed synthetic humanized insulin by joining its gene with a plasmid vector inserted into the bacterium Escherichia coli. Insulin, widely used for the treatment of diabetes, was previously extracted from the pancreas of abattoir animals (cattle and/or pigs). The resulting genetically engineered bacterium enabled the production of vast quantities of synthetic human insulin at relatively low cost. Biotechnology has also enabled emerging therapeutics like gene therapy. The application of biotechnology to basic science (for example through the Human Genome Project) has also dramatically improved our understanding of biology and as our scientific knowledge of normal and disease biology has increased, our ability to develop new medicines to treat previously untreatable diseases has increased as well.

Genetic testing allows the genetic diagnosis of vulnerabilities to inherited diseases, and can also be used to determine a child's parentage (genetic mother and father) or in general a person's ancestry. In addition to studying chromosomes to the level of individual genes, genetic testing in a broader sense includes biochemical tests for the possible presence of genetic diseases, or mutant forms of genes associated with increased risk of developing genetic disorders. Genetic testing identifies changes in chromosomes, genes, or proteins. Most of the time, testing is used to find changes that are associated with inherited disorders. The results of a genetic test can confirm or rule out a suspected genetic condition or help determine a person's chance of developing or passing on a genetic disorder. As of 2011 several hundred genetic tests were in use. Since genetic testing may open up ethical or psychological problems, genetic testing is often accompanied by genetic counseling.

Agriculture

Genetically modified crops ("GM crops", or "biotech crops") are plants used in agriculture, the DNA of which has been modified with genetic engineering techniques. In most cases the aim is to introduce a new trait to the plant which does not occur naturally in the species.

Examples in food crops include resistance to certain pests, diseases, stressful environmental conditions, resistance to chemical treatments (e.g. resistance to a herbicide), reduction of spoilage, or improving the nutrient profile of the crop. Examples in non-food crops include production of pharmaceutical agents, biofuels, and other industrially useful goods, as well as for bioremediation.

Farmers have widely adopted GM technology. Between 1996 and 2011, the total surface area of land cultivated with GM crops had increased by a factor of 94, from 17,000 square kilometers (4,200,000 acres) to 1,600,000 km (395 million acres). 10% of the world's crop lands were planted with GM crops in 2010. As of 2011, 11 different transgenic crops were grown commercially on 395 million acres (160 million hectares) in 29 countries such as the USA, Brazil, Argentina, India, Canada, China, Paraguay, Pakistan, South Africa, Uruguay, Bolivia, Australia, Philippines, Myanmar, Burkina Faso, Mexico and Spain.

Genetically modified foods are foods produced from organisms that have had specific changes introduced into their DNA with the methods of genetic engineering. These techniques have allowed for the introduction of new crop traits as well as a far greater control over a food's genetic structure than previously afforded by methods such as selective breeding and mutation breeding. Commercial sale of genetically modified foods began in 1994, when Calgene first marketed its Flavr Savr delayed ripening tomato. To date most genetic modification of foods have primarily focused on cash crops in high demand by farmers such as soybean, corn, canola, and cotton seed oil. These have been engineered for resistance to pathogens and herbicides and better nutrient profiles. GM livestock have also been experimentally developed, although as of November 2013 none are currently on the market.

There is a scientific consensus that currently available food derived from GM crops poses no greater risk to human health than conventional food, but that each GM food needs to be tested on a case-by-case basis before introduction. Nonetheless, members of the public are much less likely than scientists to perceive GM foods as safe. The legal and regulatory status of GM foods varies by country, with some nations banning or restricting them, and others permitting them with widely differing degrees of regulation.

GM crops also provide a number of ecological benefits, if not used in excess. However, opponents have objected to GM crops per se on several grounds, including environmental concerns, whether food produced from GM crops is safe, whether GM crops are needed to address the world's food needs, and economic concerns raised by the fact these organisms are subject to intellectual property law.

Industrial

Industrial biotechnology (known mainly in Europe as white biotechnology) is the application of biotechnology for industrial purposes, including industrial fermentation. It includes the practice of using cells such as micro-organisms, or components of cells like enzymes, to generate industrially useful products in sectors such as chemicals, food and feed, detergents, paper and pulp, textiles and biofuels. In doing so, biotechnology uses renewable raw materials and may contribute to lowering greenhouse gas emissions and moving away from a petrochemical-based economy.

Environmental

The environment can be affected by biotechnologies, both positively and adversely. Vallero and others have argued that the difference between beneficial biotechnology (e.g. bioremediation to clean up an oil spill or hazard chemical leak) versus the adverse effects stemming from biotechnological enterprises (e.g. flow of genetic material from transgenic organisms into wild strains) can be seen as applications and implications, respectively. Cleaning up environmental wastes is an example of an application of environmental biotechnology; whereas loss of biodiversity or loss of containment of a harmful microbe are examples of environmental implications of biotechnology.

Regulation

The regulation of genetic engineering concerns approaches taken by governments to assess and manage the risks associated with the use of genetic engineering technology, and the development and release of genetically modified organisms (GMO), including genetically modified crops and genetically modified fish. There are differences in the regulation of GMOs between countries, with some of the most marked differences occurring between the USA and Europe. Regulation varies in a given country depending on the intended use of the products of the genetic engineering. For example, a crop not intended for food use is generally not reviewed by authorities responsible for food safety. The European Union differentiates between approval for cultivation within the EU and approval for import and processing. While only a few GMOs have been approved for cultivation in the EU a number of GMOs have been approved for import and processing. The cultivation of GMOs has triggered a debate about coexistence of GM and non GM crops. Depending on the coexistence regulations incentives for cultivation of GM crops differ.

Learning

In 1988, after prompting from the United States Congress, the National Institute of General Medical Sciences (National Institutes of Health) (NIGMS) instituted a funding mechanism for biotechnology training. Universities nationwide compete for these funds to establish Biotechnology Training Programs (BTPs). Each successful application is generally funded for five years then must be competitively renewed. Graduate students in turn compete for acceptance into a BTP; if accepted, then stipend, tuition and health insurance support is provided for two or three years during the course of their Ph.D. thesis work. Nineteen institutions offer NIGMS supported BTPs. Biotechnology training is also offered at the undergraduate level and in community colleges.

Bioluminescence

Bioluminescence is the production and emission of light by a living organism. It is a form of chemiluminescence. Bioluminescence occurs widely in marine vertebrates and invertebrates, as well as in some fungi, microorganisms including some bioluminescent bacteria and terrestrial invertebrates such as fireflies. In some animals, the light is produced by symbiotic organisms such as Vibrio bacteria.

Flying and glowing firefly, Photinus pyralis

The principal chemical reaction in bioluminescence involves the light-emitting pigment luciferin and the enzyme luciferase, assisted by other proteins such as aequorin in some species. The enzyme catalyzes the oxidation of luciferin. In some species, the type of luciferin requires cofactors such as calcium or magnesium ions, and sometimes also the energy-carrying molecule adenosine triphosphate (ATP). In evolution, luciferins vary little: one in particular, coelenterazine, is found in nine different animal (phyla), though in some of these, the animals obtain it through their diet. Conversely, luciferases vary widely in different species. Bioluminescence has arisen over forty times in evolutionary history.

Female Glowworm, Lampyris noctiluca

Both Aristotle and Pliny the Elder mentioned that damp wood sometimes gives off a glow and many centuries later Robert Boyle showed that oxygen was involved in the process, both in wood and in glow-worms. It was not until the late nineteenth century that bioluminescence was properly investigated. The phenomenon is widely distributed among animal groups, especially in marine environments where dinoflagellates cause phosphorescence in the surface layers of water. On land it occurs in fungi, bacteria and some groups of invertebrates, including insects.

The uses of bioluminescence by animals include counter-illumination camouflage, mimicry of other animals, for example to lure prey, and signalling to other individuals of the same species, such as to attract mates. In the laboratory, luciferase-based systems are used in genetic engineering and for biomedical research. Other researchers are investigating the possibility of using bioluminescent systems for street and decorative lighting, and a bioluminescent plant has been created.

History

Before the development of the safety lamp for use in coal mines, dried fish skins were used in Britain and Europe as a weak source of light. This experimental form of illumination avoided the necessity of using candles which risked sparking explosions of firedamp. Another safe source of illumination in mines was bottles containing fireflies. In 1920, the American zoologist E. Newton Harvey published a monograph, The Nature of Animal Light, summarizing early work on bioluminescence. Harvey notes that Aristotle mentions light produced by dead fish and flesh, and that both Aristotle and Pliny the Elder (in his Natural History) mention light from damp wood. He also records that Robert Boyle experimented on these light sources, and showed that both they and the glow-worm require air for light to be produced. Harvey notes that in 1753, J. Baker identified the flagellate Noctiluca "as a luminous animal" "just visible to the naked eye", and in 1854 Johann Florian Heller (1813-1871) identified strands (hyphae) of fungi as the source of light in dead wood.

Tuckey, in his posthumous 1818 Narrative of the Expedition to the Zaire, described catching the animals responsible for luminescence. He mentions pellucids, crustaceans (to which he ascribes the milky whiteness of the water), and cancers (shrimps and crabs). Under the microscope he described the "luminous property" to be in the brain, resembling "a most brilliant amethyst about the size of a large pin's head".

Charles Darwin noticed bioluminescence in the sea, describing it in his Journal:

While sailing in these latitudes on one very dark night, the sea presented a wonderful and most beautiful spectacle. There was a fresh breeze, and every part of the surface, which during the day is seen as foam, now glowed with a pale light. The vessel drove before her bows two billows of liquid phosphorus, and in her wake she was followed by a milky train. As far as the eye reached, the crest of every wave was bright, and the sky above the horizon, from the reflected glare of these livid flames, was not so utterly obscure, as over the rest of the heavens.

Darwin also observed a luminous "jelly-fish of the genus Dianaea" and noted that "When the waves scintillate with bright green sparks, I believe it is generally owing to minute crustacea. But there can be no doubt that very many other pelagic animals, when alive, are phosphorescent." He guessed that "a disturbed electrical condition of the atmosphere" was probably responsible. Daniel Pauly comments that Darwin "was lucky with most of his guesses, but not here", noting that biochemistry was too little known, and that the complex evolution of the marine animals involved "would have been too much for comfort".

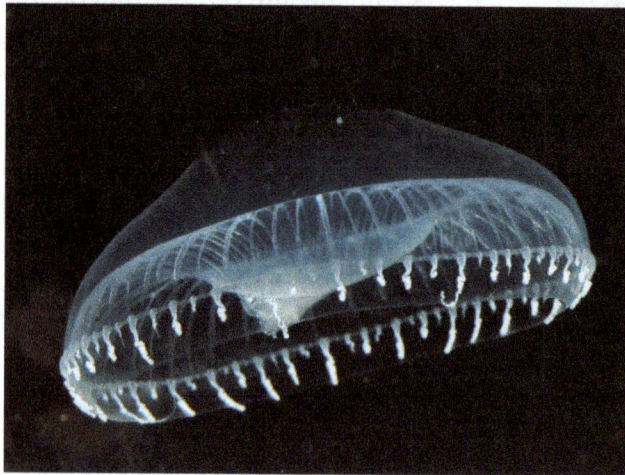

Osamu Shimomura isolated the photoprotein aequorin and its cofactor coelenterazine from the crystal jelly Aequorea victoria in 1961.

Bioluminescence attracted the attention of the United States Navy in the Cold War, since submarines in some waters can create a bright enough wake to be detected; a German submarine was sunk in the First World War, having been detected in this way. The navy was interested in predicting when such detection would be possible, and hence guiding their own submarines to avoid detection.

Among the anecdotes of navigation by bioluminescence, the Apollo 13 astronaut Jim Lovell recounted how as a navy pilot he had found his way back to his aircraft carrier USS Shangri-La when his navigation systems failed. Turning off his cabin lights, he saw the glowing wake of the ship, and was able to fly to it and land safely.

The French pharmacologist Raphaël Dubois carried out work on bioluminescence in the late nineteenth century. He studied click beetles (Pyrophorus) and the marine bivalve mollusc Pholas dactylus. He refuted the old idea that bioluminescence came from phosphorus,[a] and demonstrated that the process was related to the oxidation of a specific compound, which he named luciferin, by an enzyme. He sent Harvey siphons from the mollusc preserved in sugar. Harvey had become interested in bioluminescence as a result of visiting the South Pacific and Japan and observing phosphorescent organisms there. He studied the phenomenon for many years. His research aimed to demonstrate that luciferin, and the enzymes that act on it is to produce light, were interchangeable between species, showing that all bioluminescent organisms had a common ancestor. However, he found this hypothesis to be false, with different organisms having major differences in the composition of their light-producing proteins. He spent the next thirty years purifying and studying the components, but it fell to the young Japanese chemist Osamu Shimomura to be the first to obtain crystalline luciferin. He used the sea firefly Vargula hilgendorfii, but it was another ten years before he discovered the chemical's structure and was able to publish his 1957 paper Crystalline Cypridina Luciferin. More recently, Martin Chalfie, Osamu Shimomura and Roger Y. Tsien won the 2008 Nobel Prize in Chemistry for their 1961 discovery and development of green fluorescent protein as a tool for biological research.

Evolution

Bioluminescence in fish began at least by the Cretaceous period. About 1,500 fish species are known to be bioluminescent, and this feature evolved independently at a minimum of 27 times. Of these 27 occasions, 17 involved the taking up of bioluminous bacteria from the surrounding water while in the others, the intrinsic light evolved through chemical synthesis. These fish have become surprisingly diverse in the deep ocean and control their light with the help of their nervous system, using it not just to lure prey or hide from predators, but also for communication.

Chemical Mechanism

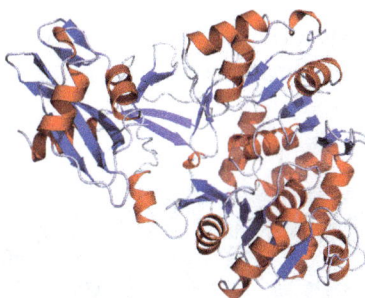

Protein folding structure of the luciferase of the firefly Photinus pyralis. The enzyme is a much larger molecule than luciferin.

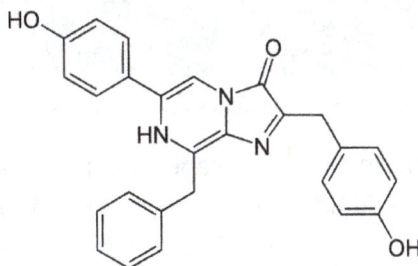

Coelenterazine is a luciferin found in many different marine phyla fromcomb jellies to vertebrates. Like all luciferins, it is oxidised to produce light.

Bioluminescence is a form of chemiluminescence where light energy is released by a chemical reaction. Fireflies, anglerfish, and other organisms produce the light-emitting pigment luciferin and the enzyme luciferase. Luciferin reacts with oxygen to create light:

Carbon dioxide (CO_2), adenosine monophosphate (AMP) and phosphate groups (PP) are released as waste products. Luciferase catalyzes the reaction, which may be mediated by cofactors such as calcium (Ca^+) or magnesium (Mg^+) ions, and for some types of luciferin (L) also the energy-carrying molecule adenosine triphosphate (ATP). The reaction can occur either inside or outside the cell. In bacteria such as Aliivibrio, the expression of genes related to bioluminescence is controlled by the lux operon.

In evolution, luciferins generally vary little: one in particular, coelenterazine, is the light emitting pigment for nine ancient phyla (groups of very different organisms), including polycystine radiolaria, Cercozoa (Phaeodaria), protozoa, comb jellies, cnidaria including jellyfish and corals, crustaceans, molluscs, arrow worms and vertebrates (ray-finned fish). Not all these organisms synthesize coelenterazine: some of them obtain it through their diet. Conversely, luciferase enzymes vary widely and tend to be different in each species. Overall, bioluminescence has arisen over forty times in evolutionary history.

Luciferin-luciferase reactions are not the only way that organisms produce light. The parchment worm Chaetopterus (a marine Polychaete) makes use of another photoprotein, aequorin, instead of luciferase. When calcium ions are added, the aequorin's rapid catalysis creates a brief flash quite unlike the prolonged glow produced by luciferase. In a second, much slower, step luciferin is regenerated from the oxidised (oxyluciferin) form, allowing it to recombine with aequorin, in readiness for a subsequent flash. Photoproteins are thus enzymes, but with unusual reaction kinetics.

In the hydrozoan jellyfish Aequorea victoria, some of the blue light released by aequorin in contact with calcium ions is absorbed by green fluorescent protein; it in turn releases green light.

Distribution

Huge numbers of bioluminescent dinoflagellates creating phosphorescence in breaking waves

Bioluminescence occurs widely among animals, especially in the open sea, including fish, jellyfish, comb jellies, crustaceans, and cephalopod molluscs; in some fungi and bacteria; and in various terrestrial invertebrates including insects. Many, perhaps most deep-sea animals produce light. Most marine light-emission is in the blue and green light spectrum. However, some loose-jawed fish emit red and infrared light, and the genus Tomopteris emits yellow light.

The most frequently encountered bioluminescent organisms may be the dinoflagellates present in the surface layers of the sea, which are responsible for the sparkling phosphorescence sometimes seen at night in disturbed water. At least eighteen genera exhibit luminosity. A different effect is the thousands of square miles of the ocean which shine with the light produced by bioluminescent bacteria, known as mareel or the milky seas effect.

Non-marine bioluminescence is less widely distributed, the two best-known cases being in fireflies and glow worms. Other invertebrates including insect larvae, annelids and arachnids possess bioluminescent abilities. Some forms of bioluminescence are brighter (or exist only) at night, following a circadian rhythm.

Uses in Nature

Bioluminescence has several functions in different taxa. Haddock et al. (2010) list as more or less definite functions in marine organisms the following: defensive functions of startle, counterillumination (camouflage), misdirection (smoke screen), distractive body parts, burglar alarm (making predators easier for higher predators to see), and warning to deter settlers; offensive functions of lure, stun or confuse prey, illuminate prey, and mate attraction/recognition. It is much easier for researchers to detect that a species is able to produce light than to analyse the chemical mechanisms or to prove what function the light serves. In some cases the function is unknown, as with species in three families of earthworm (Oligochaeta), such as Diplocardia longa where the coelomic fluid produces light when the animal moves. The following functions are reasonably well established in the named organisms.

Counterillumination Camouflage

Principle of counterillumination camouflage in firefly squid, Watasenia scintillans. When seen from below by a predator, the bioluminescence helps to match the squid's brightness and colour to the sea surface above.

In many animals of the deep sea, including several squid species, bacterial bioluminescence is used for camouflage by counterillumination, in which the animal matches the overhead environmental light as seen from below. In these animals, photoreceptors control the illumination to match the brightness of the background. These light organs are usually separate from the tissue containing the bioluminescent bacteria. However, in one species, Euprymna scolopes, the bacteria are an integral component of the animal's light organ.

Attraction

A fungus gnat from New Zealand, Arachnocampa luminosa, lives in the predator-free environment of caves and its larvae emit bluish-green light. They dangle silken threads that glow and attract flying insects, and wind in their fishing-lines when prey becomes entangled. The bioluminescence of the larvae of another fungus gnat from North America which lives on streambanks and under overhangs has a similar function. Orfelia fultoni builds sticky little webs and emits light of a deep blue colour. It has an inbuilt biological clock and, even when kept in total darkness, turns its light on and off in a circadian rhythm.

Fireflies use light to attract mates. Two systems are involved according to species; in one, females emit light from their abdomens to attract males; in the other, flying males emit signals to which the sometimes sedentary females respond. Click beetles emit an orange light from the abdomen when flying and a green light from the thorax when they are disturbed or moving about on the ground. The former is probably a sexual attractant but the latter may be defensive. Larvae of the click beetle

Pyrophorus nyctophanus live in the surface layers of termite mounds in Brazil. They light up the mounds by emitting a bright greenish glow which attracts the flying insects on which they feed.

In the marine environment, use of luminescence for mate attraction is chiefly known among ostracods, small shrimplike crustaceans, especially in the Cyprididae family. Pheromones may be used for long-distance communication, with bioluminescence used at close range to enable mates to "home in". A polychaete worm, the Bermuda fireworm creates a brief display, a few nights after the full moon, when the female lights up to attract males.

Defense

Many cephalopods, including at least 70 genera of squid, are bioluminescent. Some squid and small crustaceans use bioluminescent chemical mixtures or bacterial slurries in the same way as many squid use ink. A cloud of luminescent material is expelled, distracting or repelling a potential predator, while the animal escapes to safety. The deep sea squid Octopoteuthis deletron may autotomise portions of its arms which are luminous and continue to twitch and flash, thus distracting a predator while the animal flees.

Dinoflagellates may use bioluminescence for defence against predators. They shine when they detect a predator, possibly making the predator itself more vulnerable by attracting the attention of predators from higher trophic levels. Grazing copepods release any phytoplankton cells that flash, unharmed; if they were eaten they would make the copepods glow, attracting predators, so the phytoplankton's bioluminescence is defensive. The problem of shining stomach contents is solved (and the explanation corroborated) in predatory deep-sea fishes: their stomachs have a black lining able to keep the light from any bioluminescent fish prey which they have swallowed from attracting larger predators.

The sea-firefly is a small crustacean living in sediment. At rest it emits a dull glow but when disturbed it darts away leaving a cloud of shimmering blue light to confuse the predator. During World War II it was gathered and dried for use by the Japanese military as a source of light during clandestine operations.

The larvae of railroad worms (Phrixothrix) have paired photic organs on each body segment, able to glow with green light; these are thought to have a defensive purpose. They also have organs on the head which produce red light; they are the only terrestrial organisms to emit light of this colour.

Warning

Aposematism is a widely used function of bioluminescence, providing a warning that the creature concerned is unpalatable. It is suggested that many firefly larvae glow to repel predators; millipedes glow for the same purpose. Some marine organisms are believed to emit light for a similar reason. These include scale worms, jellyfish and brittle stars but further research is needed to fully establish the function of the luminescence. Such a mechanism would be of particular advantage to soft-bodied cnidarians if they were able to deter predation in this way. The limpet Latia neritoides is the only known freshwater gastropod that emits light. It produces greenish luminescent mucus which may have an anti-predator function. The marine snail Hinea brasiliana uses flashes of light, probably to deter predators. The blue-green light is emitted through the translucent shell, which functions as an efficient diffuser of light.

Communication

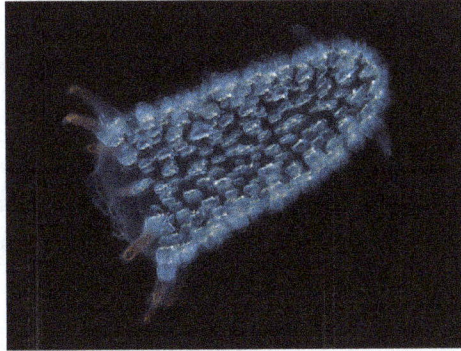

Pyrosoma, a colonial tunicate; each individual zooid in the colony flashes a blue-green light.

Communication in the form of quorum sensing plays a role in the regulation of luminescence in many species of bacteria. Small extracellularly secreted molecules stimulate the bacteria to turn on genes for light production when cell density, measured by concentration of the secreted molecules, is high.

Pyrosomes are colonial tunicates and each zooid has a pair of luminescent organs on either side of the inlet siphon. When stimulated by light, these turn on and off, causing rhythmic flashing. No neural pathway runs between the zooids, but each responds to the light produced by other individuals, and even to light from other nearby colonies. Communication by light emission between the zooids enables coordination of colony effort, for example in swimming where each zooid provides part of the propulsive force.

Some bioluminous bacteria infect nematodes that parasitize Lepidoptera larvae. When these caterpillars die, their luminosity may attract predators to the dead insect thus assisting in the dispersal of both bacteria and nematodes. A similar reason may account for the many species of fungi that emit light. Species in the genera Armillaria, Mycena, Omphalotus, Panellus, Pleurotus and others do this, emitting usually greenish light from the mycelium, cap and gills. This may attract night-flying insects and aid in spore dispersal, but other functions may also be involved.

Quantula striata is the only known bioluminescent terrestrial mollusc. Pulses of light are emitted from a gland near the front of the foot and may have a communicative function, although the adaptive significance is not fully understood.

Mimicry

A deep sea anglerfish, Bufoceratias wedli, showing the esca (lure)

Bioluminescence is used by a variety of animals to mimic other species. Many species of deep sea fish such as the anglerfish and dragonfish make use of aggressive mimicry to attract prey. They have an appendage on their heads called an esca that contains bioluminescent bacteria able to produce a long-lasting glow which the fish can control. The glowing esca is dangled or waved about to lure small animals to within striking distance of the fish.

The cookiecutter shark uses bioluminescence to camouflage its underside by counterillumination, but a small patch near its pectoral fins remains dark, appearing as a small fish to large predatory fish like tuna and mackerel swimming beneath it. When such fish approach the lure, they are bitten by the shark.

Female Photuris fireflies sometimes mimic the light pattern of another firefly, Photinus, to attract its males as prey. In this way they obtain both food and the defensive chemicals named lucibufagins, which Photuris cannot synthesize.

South American giant cockroaches of the genus Lucihormetica were believed to be the first known example of defensive mimicry, emitting light in imitation of bioluminescent, poisonous click beetles. However, doubt has been cast on this assertion, and there is no conclusive evidence that the cockroaches are bioluminescent.

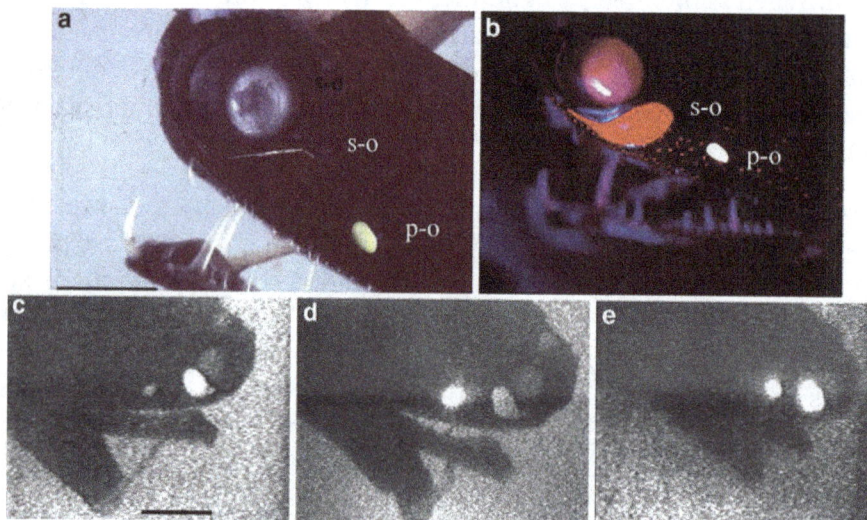

Flashing of photophores of black dragonfish, Malacosteus niger, showing red fluorescence

Illumination

While most marine bioluminescence is green to blue, some deep sea barbeled dragonfishes in the genera Aristostomias, Pachystomias and Malacosteus emit a red glow. This adaptation allows the fish to see red-pigmented prey, which are normally invisible in the deep ocean environment where red light has been filtered out by the water column.

The black dragonfish (also called the northern stoplight loosejaw) Malacosteus niger is believed to be one of the only fish to produce a red glow. Its eyes, however, are insensitive to this wavelength; it has an additional retinal pigment which fluoresces blue-green when illuminated. This alerts the fish to the presence of its prey. The additional pigment is thought to be assimilated from chlorophyll derivatives found in the copepods which form part of its diet.

Biotechnology

Biology and Medicine

Bioluminescent organisms are a target for many areas of research. Luciferase systems are widely used in genetic engineering as reporter genes, each producing a different colour by fluorescence, and for biomedical research using bioluminescence imaging. For example, the firefly luciferase gene was used as early as 1986 for research using transgenic tobacco plants. Vibrio bacteria symbiose with marine invertebrates such as the Hawaiian bobtail squid (Euprymna scolopes), are key experimental models for bioluminescence. Bioluminescent activated destruction is an experimental cancer treatment. Optogenetics which involves the use of light to control cells in living tissue, typically neurons, that have been genetically modified to express light-sensitive ion channels, a photon of non-thermal origin in the visible and ultraviolet spectrum emitted from a biological system.

Light Production

The structures of photophores, the light producing organs in bioluminescent organisms, are being investigated by industrial designers. Engineered bioluminescence could perhaps one day be used to reduce the need for street lighting, or for decorative purposes if it becomes possible to produce light that is both bright enough and can be sustained for long periods at a workable price. The gene that makes the tails of fireflies glow has been added to mustard plants. The plants glow faintly for an hour when touched, but a sensitive camera is needed to see the glow. University of Wisconsin–Madison is researching the use of genetically engineered bioluminescent E. coli bacteria, for use as bioluminescent bacteria in a light bulb. In June 2013 the Glowing Plant project raised nearly $500,000 on the crowdfunding site Kickstarter to create a bioluminescent plant. An iGEM team from Cambridge (England) has started to address the problem that luciferin is consumed in the light-producing reaction by developing a genetic biotechnology part that codes for a luciferin regenerating enzyme from the North American firefly; this enzyme "helps to strengthen and sustain light output". In 2016, Glowee, a French company started selling bioluminescent lights, tageting shop fronts and municipal street signs as their main markets. France has a law that forbids retailers and offices from illumunating their windows between 1 and 7 in the morning in order to minimise energy consumption and pollution. Glowee hoped their product would get round this ban. They used bacteria called Aliivibrio fischeri which glow in the dark but the maximum lifetime of their product was three days.

References

- Smiles, Samuel (1862). Lives of the Engineers. Volume III (George and Robert Stephenson). London: John Murray. p. 107. ISBN 0-7153-4281-9. (ISBN refers to the David & Charles reprint of 1968 with an introduction by L. T. C. Rolt)

- Pauly, Daniel (13 May 2004). Darwin's Fishes: An Encyclopedia of Ichthyology, Ecology, and Evolution. Cambridge University Press. pp. 15–16. ISBN 978-1-139-45181-9.

- Huth, John Edward (15 May 2013). The Lost Art of Finding Our Way. Harvard University Press. p. 423. ISBN 978-0-674-07282-4.

- Pieribone, Vincent; Gruber, David F. (2005). Aglow in the Dark: The Revolutionary Science of Biofluorescence. Harvard University Press. pp. 35–41. ISBN 978-0-674-01921-8.

- Davis, Matthew P.; Sparks, John S.; Smith, W. Leo (2016-06-08). "Repeated and Widespread Evolution of Bioluminescence in Marine Fishes". PLOS ONE 11 (6): e0155154. doi:10.1371/journal.pone.0155154. ISSN 1932-6203.

- Marcellin, Frances (26 February 2016). "Glow-in-the-dark bacterial lights could illuminate shop windows". The New Scientist. Retrieved 4 March 2016.

- Tuckey, James Hingston (May 1818). Thomson, Thomas, ed. Narrative of the Expedition to the Zaire. Annals of Philosophy. volume XI. p. 392. Retrieved 22 April 2015.

- Branham, Marc. "Glow-worms, railroad-worms (Insecta: Coleoptera: Phengodidae)". Featured Creatures. University of Florida. Retrieved 29 November 2014.

- Nordgren, I. K.; Tavassoli, A. (2014). "A bidirectional fluorescent two-hybrid system for monitoring protein-protein interactions". Molecular BioSystems 10 (3): 485–490. doi:10.1039/c3mb70438f. PMID 24382456.

- Ludwig Institute for Cancer Research (21 April 2003). "Firefly Light Helps Destroy Cancer Cells; Researchers Find That The Bioluminescence Effects Of Fireflies May Kill Cancer Cells From Within". Science Daily. Retrieved 4 December 2014.

- Reshetiloff, Kathy (1 July 2001). "Chesapeake Bay night-lights add sparkle to woods, water". Bay Journal. Retrieved 16 December 2014.

- Greven, Hartmut; Zwanzig, Nadine (2013). "Courtship, Mating, and Organisation of the Pronotum in the Glow-spot Cockroach Lucihormetica verrucosa (Brunner von Wattenwyl, 1865) (Blattodea: Blaberidae)". Entomologie heute 25: 77–9.

Evolution of Biochemistry

This chapter provides thorough knowledge about the history, evolution and current importance of biochemistry. It provides information about enzymes and metabolism. The chapter aims to elaborate the evolution of biochemistry since its early stages to becoming a specialized field in the modern times.

The history of biochemistry can be said to have started with the ancient Greeks who were interested in the composition and processes of life, although biochemistry as a specific scientific discipline has its beginning around the early 19th century. Some argued that the beginning of biochemistry may have been the discovery of the first enzyme, diastase (today called amylase), in 1833 by Anselme Payen, while others considered Eduard Buchner's first demonstration of a complex biochemical process alcoholic fermentation in cell-free extracts to be the birth of biochemistry. Some might also point to the influential work of Justus von Liebig from 1842, Animal chemistry, or, Organic chemistry in its applications to physiology and pathology, which presented a chemical theory of metabolism, or even earlier to the 18th century studies on fermentation and respiration by Antoine Lavoisier.

The term "biochemistry" itself is derived from the combining form bio-, meaning "life", and chemistry. The word is first recorded in English in 1848, while in 1877, Felix Hoppe-Seyler used the term (Biochemie in German) in the foreword to the first issue of Zeitschrift für Physiologische Chemie (Journal of Physiological Chemistry) as a synonym for physiological chemistry and argued for the setting up of institutes dedicate to its studies. Nevertheless, several sources cite German chemist Carl Neuberg as having coined the term for the new discipline in 1903, and some credit it to Franz Hofmeister.

The subject of study in biochemistry is the chemical processes in living organisms, and its history involves the discovery and understanding of the complex components of life and the elucidation of pathways of biochemical processes. Much of biochemistry deals with the structures and functions of cellular components such as proteins, carbohydrates, lipids, nucleic acids and other biomolecules; their metabolic pathways and flow of chemical energy through metabolism; how biological molecules give rise to the processes that occur within living cells; it also focuses on the biochemical processes involved in the control of information flow through biochemical signalling, and how they relate to the functioning of whole organisms. Over the last 40 years the field has had success in explaining living processes such that now almost all areas of the life sciences from botany to medicine are engaged in biochemical research.

Among the vast number of different biomolecules, many are complex and large molecules (called polymers), which are composed of similar repeating subunits (called monomers). Each class of polymeric biomolecule has a different set of subunit types. For example, a protein is a polymer whose subunits are selected from a set of twenty or more amino acids, carbohydrates are formed from sugars known as monosaccharides, oligosaccharides, and polysaccharides, lipids are formed from fatty acids and glycerols, and nucleic acids are formed from nucleotides. Biochemistry studies the chemical properties of important biological molecules, like proteins, and in particular the chemistry of enzyme-catalyzed reactions. The biochemistry of cell metabolism and the endocrine system has been extensively described. Other areas of biochemistry include the genetic code (DNA, RNA), protein synthesis, cell membrane transport, and signal transduction.

Protobiochemistry

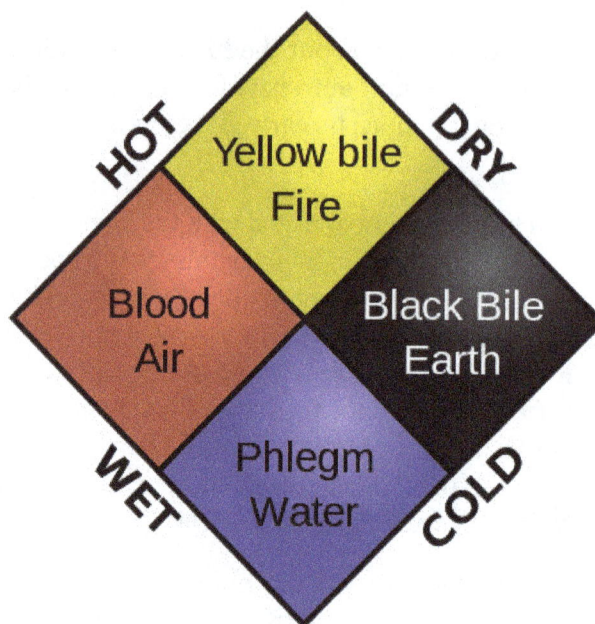

In this diagram, each kind of food would result in a different physiological result. For example, cold and dry food would produce black bile.

In these regards, the study of biochemistry began when biology first began to interest society—as the ancient Chinese developed a system of medicine based on yin and yang, and also the five phases, which both resulted from alchemical and biological interests. It began in the ancient Indian culture also with an interest in medicine, as they developed the concept of three humors that were similar to the Greek's four humours. They also delved into the interest of bodies being composed of tissues. As in the majority of early sciences, the Islamic world greatly contributed to early biological advancements as well as alchemical advancements; especially with the introduction of clinical trials and clinical pharmacology presented in Avicenna's The Canon of Medicine. On the side of chemistry, early advancements were heavily attributed to exploration of alchemical interests but also included: metallurgy, the scientific method, and early theories of atomism. In more recent times, the study of chemistry was marked by milestones such as the development of Mendeleev's periodic table, Dalton's atomic model, and the conservation of mass theory. This last mention has the most importance of the three due to the fact that this law intertwines chemistry with thermodynamics in an intercalated manner.

Enzymes

As early as the late 18th century and early 19th century, the digestion of meat by stomach secretions and the conversion of starch to sugars by plant extracts and saliva were known. However, the mechanism by which this occurred had not been identified.

In the 19th century, when studying the fermentation of sugar to alcohol by yeast, Louis Pasteur concluded that this fermentation was catalyzed by a vital force contained within the yeast cells called ferments, which he thought functioned only within living organisms. He wrote that "alcoholic fermentation is an act correlated with the life and organization of the yeast cells, not with the death or putrefaction of the cells."

Anselme Payen discovered in 1833 the first enzyme who called diastase and in 1878 German physiologist Wilhelm Kühne (1837–1900) coined the term enzyme, to describe this process. The word enzyme was used later to refer to nonliving substances such as pepsin, and the word ferment used to refer to chemical activity produced by living organisms.

In 1897 Eduard Buchner began to study the ability of yeast extracts to ferment sugar despite the absence of living yeast cells. In a series of experiments at the University of Berlin, he found that the sugar was fermented even when there were no living yeast cells in the mixture. He named the enzyme that brought about the fermentation of sucrose "zymase". In 1907 he received the Nobel Prize in Chemistry "for his biochemical research and his discovery of cell-free fermentation". Following Buchner's example; enzymes are usually named according to the reaction they carry out. Typically the suffix -ase is added to the name of the substrate (e.g., lactase is the enzyme that cleaves lactose) or the type of reaction (e.g., DNA polymerase forms DNA polymers).

Having shown that enzymes could function outside a living cell, the next step was to determine their biochemical nature. Many early workers noted that enzymatic activity was associated with proteins, but several scientists (such as Nobel laureate Richard Willstätter) argued that proteins were merely carriers for the true enzymes and that proteins per se were incapable of catalysis. However, in 1926, James B. Sumner showed that the enzyme urease was a pure protein and crystallized it; Sumner did likewise for the enzyme catalase in 1937. The conclusion that pure proteins can be enzymes was definitively proved by Northrop and Stanley, who worked on the digestive enzymes pepsin (1930), trypsin and chymotrypsin. These three scientists were awarded the 1946 Nobel Prize in Chemistry.

This discovery, that enzymes could be crystallized, meant that scientists eventually could solve their structures by x-ray crystallography. This was first done for lysozyme, an enzyme found in tears, saliva and egg whites that digests the coating of some bacteria; the structure was solved by a group led by David Chilton Phillips and published in 1965. This high-resolution structure of lysozyme marked the beginning of the field of structural biology and the effort to understand how enzymes work at an atomic level of detail.

Metabolism

Early Metabolic Interest

The term metabolism is derived – Metabolismos for "change", or "overthrow". The history of the scientific study of metabolism spans 800 years. The earliest of all metabolic studies began during the early thirteenth century (1213-1288) by a Muslim scholar from Damascus named Ibn al-Nafis. al-Nafis stated in his most well-known work Theologus Autodidactus that "that body and all its parts are in a continuous state of dissolution and nourishment, so they are inevitably undergoing permanent change." Although al-Nafis was the first documented physician to have an interest in biochemical concepts, the first controlled experiments in human metabolism were published by Santorio Santorio in 1614 in his book Ars de statica medecina. This book describes how he weighed himself before and after eating, sleeping, working, sex, fasting, drinking, and excreting. He found that most of the food he took in was lost through what he called "insensible perspiration".

Metabolism: 20Th Century - Present

One of the most prolific of these modern biochemists was Hans Krebs who made huge contributions to the study of metabolism. He discovered the urea cycle and later, working with Hans Kornberg, the citric acid cycle and the glyoxylate cycle. These discoveries led to Krebs being awarded the Nobel Prize in physiology in 1953, which was shared with the German biochemist Fritz Albert Lipmann who also codiscovered the essential cofactor coenzyme A.

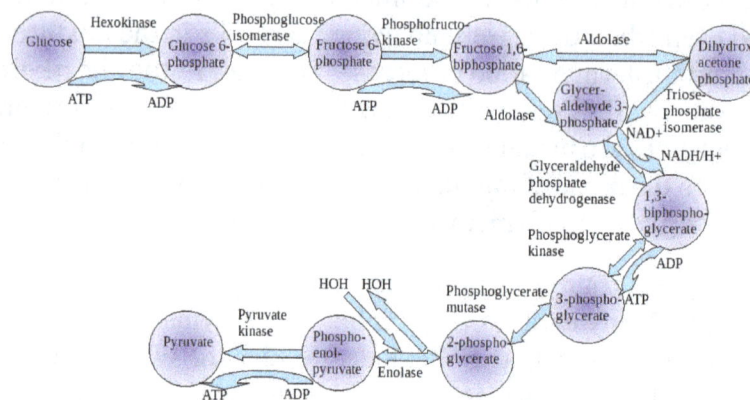

Shown here is a step-wise depiction of glycolysis along with the required enzymes.

Glucose Absorption

In 1960, the biochemist Robert K. Crane revealed his discovery of the sodium-glucose cotransport as the mechanism for intestinal glucose absorption. This was the very first proposal of a coupling between the fluxes of an ion and a substrate that has been seen as sparking a revolution in biology. This discovery, however, would not have been possible if it were not for the discovery of the molecule glucose's structure and chemical makeup. These discoveries are largely attributed to the German chemist Emil Fischer who received the Nobel Prize in chemistry nearly 60 years earlier.

Glycolysis

Since metabolism focuses on the breaking down (catabolic processes) of molecules and the building of larger molecules from these particles (anabolic processes), the use of glucose and its involvement in the formation of adenosine triphosphate (ATP) is fundamental to this understanding. The most frequent type of glycolysis found in the body is the type that follows the Embden-Meyerhof-Parnas (EMP) Pathway, which was discovered by Gustav Embden, Otto Meyerhof, and Jakob Karol Parnas. These three men discovered that glycolysis is a strongly determinant process for the efficiency and production of the human body. The significance of the pathway shown in the image to the right is that by identifying the individual steps in this process doctors and researchers are able to pinpoint sites of metabolic malfunctions such as pyruvate kinase deficiency that can lead to severe anemia. This is most important because cells, and therefore organisms, are not capable of surviving without proper functioning metabolic pathways.

Instrumental Advancements (20Th Century)

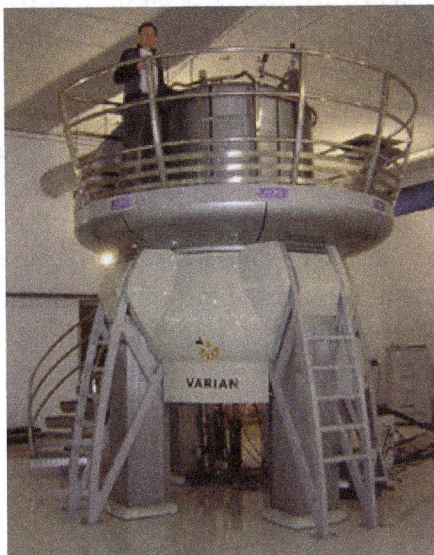

This is an example of a very large NMR instrument known as the HWB-NMR with a 21.2T (Tesla) magnet.

Since then, biochemistry has advanced, especially since the mid-20th century, with the development of new techniques such as chromatography, X-ray diffraction, NMR spectroscopy, radioisotopic labelling, electron microscopy and molecular dynamics simulations. These techniques allowed for the discovery and detailed analysis of many molecules and metabolic pathways of the cell, such as glycolysis and the Krebs cycle (citric acid cycle). The example of an NMR instrument shows that some of these instruments, such as the HWB-NMR, can be very large in size and can cost anywhere from a few hundred dollars to millions of dollars ($16 million for the one shown here).

Polymerase Chain Reaction

Shown above is a model of a thermo cycler that is currently being used in polymerase chain reaction.

Polymerase chain reaction (PCR) is the primary gene amplification technique that has revolutionized modern biochemistry. Polymerase chain reaction was developed by Kary Mullis in 1983. There are four steps to a proper polymerase chain reaction: 1) denaturation 2) extension 3) insertion (of gene to be expressed) and finally 4) amplification of the inserted gene. These steps with simple illustrative examples of this process can be seen in the image below and to the right of this section. This technique allows for the copy of a single gene to be amplified into hundreds or even millions of copies and has become a cornerstone in the protocol for any biochemist that wishes to work with bacteria and gene expression. PCR is not only used for gene expression research but is also capable of aiding laboratories in diagnosing certain diseases such a lymphomas, some types of leukemia, and other malignant diseases that can sometimes puzzle doctors. Without polymerase chain reaction development, there are many advancements in the field of bacterial study and protein expression study that would not have come to fruition. The development of the theory and process of polymerase chain reaction is essential but the invention of the thermal cycler is equally as important because the process would not be possible without this instrument. This is yet another testament to the fact that the advancement of technology is just as crucial to sciences such as biochemistry as is the painstaking research that leads to the development of theoretical concepts.

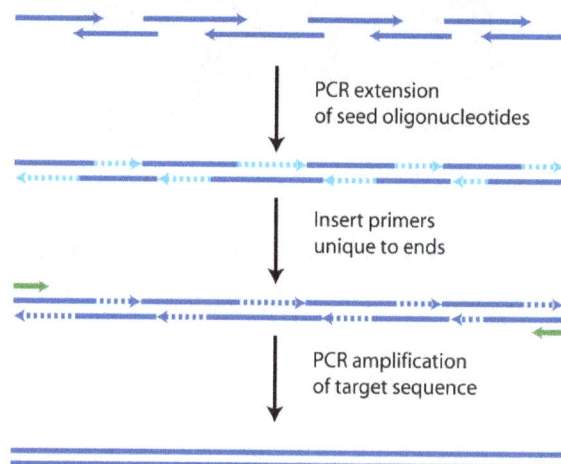

PCR extension of seed oligonucleotides

Insert primers unique to ends

PCR amplification of target sequence

Shown here are the three steps of PCR, following the first step of denaturation.

References

- Clarence Peter Berg (1980). "The University of Iowa and Biochemistry from Their Beginnings": 1–2. ISBN 9780874140149.

- Frederic Lawrence Holmes (1987). Lavoisier and the Chemistry of Life: An Exploration of Scientific Creativity. University of Wisconsin Press. p. xv. ISBN 978-0299099848.

- Anne-Katrin Ziesak; Hans-Robert Cram (18 October 1999). Walter de Gruyter Publishers, 1749-1999. Walter de Gruyter & Co. p. 169. ISBN 978-3110167412.

- Horst Kleinkauf, Hans von Döhren, Lothar Jaenicke (1988). The Roots of Modern Biochemistry: Fritz Lippmann's Squiggle and its Consequences. Walter de Gruyter & Co. p. 116. ISBN 9783110852455.

- Mark Amsler (1986). The Languages of Creativity: Models, Problem-solving, Discourse. University of Delaware Press. p. 55. ISBN 978-0874132809.

- Advances in Carbohydrate Chemistry and Biochemistry, Volume 70. Academic Press. 28 November 2013. p. 36. ASIN B00H7E78BG.

Permissions

We would like to thank the editorial team for lending their expertise to make the book truly unique. They have played a crucial role in the development of this book. Without their invaluable contributions this book wouldn't have been possible. They have made vital efforts to compile up to date information on the varied aspects of this subject to make this book a valuable addition to the collection of many professionals and students.

This book was conceptualized with the vision of imparting up-to-date and integrated information in this field. To ensure the same, a matchless editorial board was set up. Every individual on the board went through rigorous rounds of assessment to prove their worth. After which they invested a large part of their time researching and compiling the most relevant data for our readers.

The editorial board has been involved in producing this book since its inception. They have spent rigorous hours researching and exploring the diverse topics which have resulted in the successful publishing of this book. They have passed on their knowledge of decades through this book. To expedite this challenging task, the publisher supported the team at every step. A small team of assistant editors was also appointed to further simplify the editing procedure and attain best results for the readers.

Apart from the editorial board, the designing team has also invested a significant amount of their time in understanding the subject and creating the most relevant covers. They scrutinized every image to scout for the most suitable representation of the subject and create an appropriate cover for the book.

The publishing team has been an ardent support to the editorial, designing and production team. Their endless efforts to recruit the best for this project, has resulted in the accomplishment of this book. They are a veteran in the field of academics and their pool of knowledge is as vast as their experience in printing. Their expertise and guidance has proved useful at every step. Their uncompromising quality standards have made this book an exceptional effort. Their encouragement from time to time has been an inspiration for everyone.

The publisher and the editorial board hope that this book will prove to be a valuable piece of knowledge for students, practitioners and scholars across the globe.

Index

www.ingramcontent.com/pod-product-compliance
Lightning Source LLC
Chambersburg PA
CBHW082030190326
41458CB00010B/3323